I0475496

IN THE SAME SERIES.

ON THE STUDY AND DIFFICULTIES OF MATHE-
MATICS. By AUGUSTUS DE MORGAN. Entirely new edition, with portrait of the author, index, and annotations,
bibliographies of modern works on algebra, the philosophy
of mathematics, pan-geometry, etc. Pp., 288. Cloth, $1.25
net (5s.).

LECTURES ON ELEMENTARY MATHEMATICS. By
JOSEPH LOUIS LAGRANGE. Translated from the French by
Thomas J. McCormack. With photogravure portrait of
Lagrange, notes, biography, marginal analyses, etc. Only
separate edition in French or English. Pages, 172. Cloth,
$1.00 net (5s.).

ELEMENTARY ILLUSTRATIONS OF THE DIFFEREN-
TIAL AND INTEGRAL CALCULUS. By AUGUSTUS DE
MORGAN. New reprint edition. With sub-headings, and
a brief bibliography of English, French, and German text-
books of the Calculus. Pp., 144. Price, $1.00 net (5s.).

MATHEMATICAL ESSAYS AND RECREATIONS. By
HERMANN SCHUBERT, Professor of Mathematics in the
Johanneum, Hamburg, Germany. Translated from the
German by *Thomas J. McCormack.* Containing essays on
The Notion and Definition of Number, Monism in Arith-
metic, The Nature of Mathematical Knowledge, The
Magic Square, The Fourth Dimension, The Squaring of
the Circle. Pages, 149. Cuts, 37. Price, Cloth, 75c net
(3s. 6d.).

A BRIEF HISTORY OF ELEMENTARY MATHEMATICS.
By DR. KARL FINK, late Professor in Tübingen. Translated
from the German by Prof. *Wooster Woodruff Beman* and
Prof. *David Eugene Smith.* Pp. 333. Price, cloth, $1.50
net (6s.).

THE OPEN COURT PUBLISHING CO.
324 DEARBORN ST., CHICAGO.

A BRIEF

HISTORY OF MATHEMATICS

AN AUTHORIZED TRANSLATION OF

DR. KARL FINK'S GESCHICHTE DER
ELEMENTAR-MATHEMATIK

BY

WOOSTER WOODRUFF BEMAN
PROFESSOR OF MATHEMATICS IN THE UNIVERSITY OF MICHIGAN

AND

DAVID EUGENE SMITH
PRINCIPAL OF THE STATE NORMAL SCHOOL AT BROCKPORT, N. Y.

***

CHICAGO
THE OPEN COURT PUBLISHING COMPANY

LONDON AGENTS
KEGAN PAUL, TRENCH, TRÜBNER & CO., LTD.
1900

TRANSLATION COPYRIGHTED
BY
THE OPEN COURT PUBLISHING CO.
1900.

TRANSLATORS' PREFACE.

THE translators feel that no apology is necessary for any reasonable effort to encourage the study of the history of mathematics. The clearer view of the science thus afforded the teacher, the inspiration to improve his methods of presenting it, the increased interest in the class-work, the tendency of the subject to combat stagnation of curricula,—these are a few of the reasons for approving the present renaissance of the study.

This phase of scientific history which Montucla brought into such repute—it must be confessed rather by his literary style than by his exactness—and which writers like De Morgan in England, Chasles in France, Quetelet in Belgium, Hankel and Baltzer in Germany, and Boncompagni in Italy encouraged as the century wore on, is seeing a great revival in our day. This new movement is headed by such scholars as Günther, Eneström, Loria, Paul Tannery, and Zeuthen, but especially by Moritz Cantor, whose *Vorlesungen über Geschichte der Mathematik* must long remain the world's standard.

In any movement of this kind compendia are always necessary for those who lack either the time or the linguistic power to read the leading treatises. Several such works have recently appeared in various languages. But the most systematic attempt in this direction is the work here translated. The writers of most handbooks of this kind feel called upon to collect a store of anecdotes, to incorporate tales of no historic value, and to minimize the real history of the science. Fink, on the other hand, omits biography entirely, referring the reader to a brief table in the appendix or to the encyclopedias. He systematically considers the growth of arithmetic, algebra, geometry, and trigonometry, carrying the historic development, as should be done, somewhat beyond the limits of the ordinary course.

At the best, the work of the translator is a rather thankless task. It is a target for critics of style and for critics of matter. For the style of the German work the translators will hardly be held responsible. It is not a fluent one, leaning too much to the scientific side to make it always easy reading. Were the work less scientific, it would lend itself more readily to a better English form, but the translators have preferred to err on the side of a rather strict adherence to the original.

As to the matter, it has seemed unwise to make any considerable changes. The attempt has been made to correct a number of unquestionable errors, occasional references have been added, and the biographical notes have been rewritten. It has not seemed advisable, however, to insert a large number of bibliographical notes. Readers who are interested in the subject will naturally place upon their shelves the works of De Morgan, Allman, Gow, Ball, Heath, and other English writers, and, as far as may be, works in other languages. The leading German authorities are mentioned in the footnotes, and the French language offers little at present beyond the works of Chasles and Paul Tannery.

The translators desire to express their obligations to Professor Markley for valuable assistance in the translation.

Inasmuch as the original title of the work, *Geschichte der Elementar-Mathematik*, is misleading, at least to English readers, the work going considerably beyond the limits of the elements, it has been thought best to use as the English title, *A Brief History of Mathematics*.

W. W. BEMAN, Ann Arbor, Mich.
D. E. SMITH, Brockport, N. Y.

March, 1900.

PREFACE.

IF the history of a science possesses value for every one whom calling or inclination brings into closer relations to it,—if the knowledge of this history is imperative for all who have influence in the further development of scientific principles or the methods of employing them to advantage, then acquaintance with the rise and growth of a branch of science is especially important to the man who wishes to teach the elements of this science or to penetrate as a student into its higher realms.

The following history of elementary mathematics is intended to give students of mathematics an historical survey of the elementary parts of this science and to furnish the teacher of the elements opportunity, with little expenditure of time, to review connectedly points for the most part long familiar to him and to utilise them in his teaching in suitable comments. The enlivening influence of historical remarks upon this elementary instruction has never been disputed. Indeed there are text-books for the elements of mathematics (among the more recent those of Baltzer and Schubert) which devote considerable space to the history of the science in the way of special notes. It is certainly desirable that instead of scattered historical references there should be offered a connected presentation of the history of elementary mathematics, not one intended for the use of scholars, not as an equivalent for the great works upon the history of mathematics, but only as a first picture, with fundamental tones clearly sustained, of the principal results of the investigation of mathematical history.

In this book the attempt has been made to differentiate the histories of the separate branches of mathematical science. There

CONTENTS.

IV. GEOMETRY.

V. TRIGONOMETRY.

GENERAL SURVEY.

THE beginnings of the development of mathematical truths date back to the earliest civilizations of which any literary remains have come down to us, namely the Egyptian and the Babylonian. On the one hand, brought about by the demands of practical life, on the other springing from the real scientific spirit of separate groups of men, especially of the priestly caste, arithmetic and geometric notions came into being. Rarely, however, was this knowledge transmitted through writing, so that of the Babylonian civilization we possess only a few traces. From the ancient Egyptian, however, we have at least one manual, that of Ahmes, which in all probability appeared nearly two thousand years before Christ.

The real development of mathematical knowledge, obviously stimulated by Egyptian and Babylonian influences, begins in Greece. This development shows itself predominantly in the realm of geometry, and enters upon its first classic period, a period of no great duration, during the era of Euclid, Archimedes, Eratosthenes, and Apollonius. Subsequently it inclines more toward the arithmetic side; but it soon becomes so completely engulfed by the heavy waves of stormy periods that only after long centuries and in a foreign soil, out of Greek works which had escaped the general destruction, could a seed, new and full of promise, take root.

One would naturally expect to find the Romans entering with eagerness upon the rich intellectual inheritance which came to them from the conquered Greeks, and to find their sons, who so willingly resorted to Hellenic masters, showing an enthusiasm for Greek mathematics. Of this, however, we have scarcely any evidence. The Romans understood very well the practical value to the statesman of Greek geometry and surveying—a thing which shows itself also in the later Greek schools—but no real mathematical advance is to be found anywhere in Roman history. Indeed, the Romans often had so mistaken an idea of Greek learning that not infrequently they handed it down to later generations in a form entirely distorted.

More important for the further development of mathematics are the relations of the Greek teachings to the investigations of the Hindus and the Arabs. The Hindus distinguished themselves by a pronounced talent for numerical calculation. What especially distinguishes them is their susceptibility to the influence of Western science, the Babylonian and especially the Greek, so that they incorporated into their own system what they received from outside sources and then worked out independent results.

The Arabs, however, in general do not show this same independence of apprehension and of judgment. Their chief merit, none the less a real one however, lies in the untiring industry which they showed in translating into their own language the literary treasures of the Hindus, Persians and Greeks. The courts of the Mohammedan princes from the ninth to the thirteenth centuries were the seats of a remarkable scientific activity, and to this circumstance alone do we owe it that after a period of long and dense darkness Western Europe was in a comparatively short time opened up to the mathematical sciences.

The learning of the cloisters in the earlier part of the Middle Ages was not by nature adapted to enter seriously into matters mathematical or to search for trustworthy sources of such knowledge. It was the Italian merchants whose practical turn and easy adaptability first found, in their commercial relations with Mohammedan West Africa and Southern Spain, abundant use for the common calculations of arithmetic. Nor was it long after that there developed among them a real spirit of discovery, and the first great triumph of the newly revived science was the solution of the cubic equation by Tartaglia. It should be said, however, that the later cloister cult labored zealously to extend the Western Arab learning by means of translations into the Latin.

In the fifteenth century, in the persons of Peurbach and Regiomontanus, Germany first took position in the great rivalry for the advancement of mathematics. From that time until the middle of the seventeenth century the German mathematicians were chiefly calculators, that is teachers in the reckoning schools (*Rechenschulen*). Others, however, were algebraists, and the fact is deserving of emphasis that there were intellects striving to reach still loftier heights. Among them Kepler stands forth pre-eminent, but with him are associated Stifel, Rudolff, and Bürgi. Certain is it that at this time and on German soil elementary arithmetic and common algebra, vitally influenced by the Italian school, attained a standing very conducive to subsequent progress.

The modern period in the history of mathematics begins about the middle of the seventeenth century. Descartes projects the foundation theory of the analytic geometry. Leibnitz and Newton appear as the discoverers of the differential calculus. The time has now come when geometry, a science only rarely, and even then but imperfectly, appreciated after its banishment from Greece, enters along with analysis upon a period of prosperous advance, and takes full advantage of this latter sister science in attaining its results. Thus there were periods in which geometry was able through its brilliant discoveries to cast analysis, temporarily at least, into the shade.

The unprecedented activity of the great Gauss divides the modern period into two parts: before Gauss—the establishment of the methods of the dif-

ferential and integral calculus and of analytic geometry as well as more restricted preparations for later
advance; with Gauss and after him—the magnificent
development of modern mathematics with its special
regions of grandeur and depth previously undreamed
of. The mathematicians of the nineteenth century
are devoting themselves to the theory of numbers,
modern algebra, the theory of functions and projective geometry, and in obedience to the impulse of
human knowledge are endeavoring to carry their light
into remote realms which till now have remained in
darkness.

I. NUMBER-SYSTEMS AND NUMBER-SYMBOLS.

AN inexhaustible profusion of external influences upon the human mind has found its legitimate expression in the formation of speech and writing in numbers and number-symbols. It is true that a counting of a certain kind is found among peoples of a low grade of civilization and even among the lower animals. "Even ducks can count their young."* But where the nature and the condition of the objects have been of no consequence in the formation of the number itself, there human counting has first begun.

The oldest counting was even in its origin a process of reckoning, an adjoining, possibly also in special elementary cases a multiplication, performed upon the objects counted or upon other objects easily employed, such as pebbles, shells, fingers. Hence arose number-names. The most common of these undoubtedly belong to the primitive domain of language ; with the advancing development of language their aggregate was gradually enlarged, the legitimate combina-

---

* Hankel, *Zur Geschichte der Mathematik im Altertum und Mittelalter*, 1874, p. 7. Hereafter referred to as Hankel. Tylor's *Primitive Culture* also has a valuable chapter upon counting.

tion of single terms permitting and favoring the creation of new numbers.　Hence arose number-systems.

The explanation of the fact that 10 is almost everywhere found as the base of the system of counting is
seen in the common use of the fingers in elementary
calculations.　In all ancient civilizations finger-reckoning was known and even to-day it is carried on to a
remarkable extent among many savage peoples.　Certain South African races use three persons for numbers which run above 100, the first counting the units
on his fingers, the second the tens, and the third the
hundreds.　They always begin with the little finger of
the left hand and count to the little finger of the right.
The first counts continuously, the others raising a
finger every time a ten or a hundred is reached.[*]

Some languages contain words belonging fundamentally to the scale of 5 or 20 without these systems
having been completely elaborated ; only in certain
places do they burst the bounds of the decimal system.　In other cases, answering to special needs, 12
and 60 appear as bases.　The New Zealanders have
a scale of 11, their language possessing words for the
first few powers of 11, and consequently 12 is represented as 11 and 1, 13 as 11 and 2, 22 as two 11's,
and so on.[†]

[*] Cantor, M., *Vorlesungen über Geschichte der Mathematik*.　Vol. I, 1880;
2nd ed., 1894, p. 6.　Hereafter referred to as Cantor.　Conant, L. L., *The Number Concept*, N. Y. 1896.　Gow, J., *History of Greek Geometry*, Cambridge, 1884,
Chap. I.

[†] Cantor, I., p. 10.

In the verbal formation of a number-system addition and multiplication stand out prominently as definitive operations for the composition of numbers; very rarely does subtraction come into use and still more rarely division.   For example, 18 is called in Latin $10 + 8$ (*decem et octo*), in Greek $8 + 10$ (ὀκτω-καί-δεκα), in French 10 8 (*dix-huit*), in German 8 10 (*acht-zehn*), in Latin also $20 - 2$ (*duo-de-viginti*), in Lower Breton $3 \cdot 6$ (*tri-omc'h*), in Welsh $2 \cdot 9$ (*dew-naw*), in Aztec $15 + 3$ (*caxtulli-om-ey*), while 50 is called in the Basque half-hundred, in Danish two-and-a-half times twenty.[*] In spite of the greatest diversity of forms, the written representation of numbers, when not confined to the mere rudiments, shows a general law according to which the higher order precedes the lower in the direction of the writing.[†]   Thus in a four-figure number the thousands are written by the Phœnicians at the right, by the Chinese above, the former writing from right to left, the latter from above downward.   A striking exception to this law is seen in the subtractive principle of the Romans in IV, IX, XL, etc., where the smaller number is written before the larger.

Among the Egyptians we have numbers running from right to left in the hieratic writing, with varying direction in the hieroglyphics.   In the latter the numbers were either written out in words or represented by symbols for each unit, repeated as often as neces-

* Hankel, p. 22.   † Hankel, p. 32.

sary. In one of the tombs near the pyramids of Gizeh have been found hieroglyphic numerals in which 1 is represented by a vertical line, 10 by a kind of horse-shoe, 100 by a short spiral, 10 000 by a pointing finger, 100 000 by a frog, 1 000 000 by a man in the attitude of astonishment.   In the hieratic symbols the figure for the unit of higher order stands to the right of the one of lower order in accordance with the law of sequence already mentioned.   The repetition of symbols for a unit of any particular order does not obtain, because there are special characters for all nine units, all the tens, all the hundreds, and all the thousands.*
We give below a few characteristic specimens of the hieratic symbols:

I	II	III	—	ꟿ	Λ	Λ̄	ꟼ	÷
1	2	3	4	5	10	20	30	40

The Babylonian cuneiform inscriptions† proceed from left to right, which must be looked upon as exceptional in a Semitic language.   In accordance with the law of sequence the units of higher order stand on the left of those of lower order.   The symbols used in writing are chiefly the horizontal wedge ➤, the vertical wedge Y, and the combination of the two at an angle ◀. The symbols were written beside one another, or, for ease of reading and to save space, over one another.   The symbols for 1, 4, 10, 100, 14, 400, respectively, are as follows:

* Cantor, I., pp. 43. 44.   † Cantor, I., pp. 77, 78.

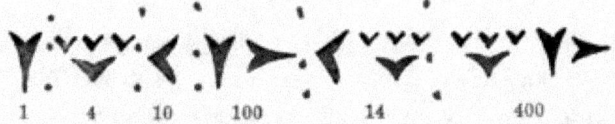

For numbers exceeding 100 there was also, besides
the mere juxtaposition, a multiplicative principle;
the symbol representing the number of hundreds was
placed at the left of the symbol for hundreds as in the
case of 400 already shown. The Babylonians probably
had no symbol for zero.* The sexagesimal system
(i. e., with the base 60), which played such a part in
the writings of the Babylonian scholars (astronomers
and mathematicians), will be mentioned later.

The Phœnicians, whose twenty-two letters were
derived from the hieratic characters of the Egyptians,
either wrote the numbers out in words or used special
numerical symbols—for the units vertical marks, for
the tens horizontal.† Somewhat later the Syrians used
the twenty-two letters of their alphabet to represent
the numbers 1, 2, . . 9, 10, 20, . . . 90, 100, . . . 400;
500 was 400 + 100, etc. The thousands were repre
sented by the symbols for units with a subscript
comma at the right.‡ The Hebrew notation follows
the same plan.

The oldest Greek numerals (aside from the written
words) were, in general, the initial letters of the funda
mental numbers. I for 1, Π for 5 (πέντε), Δ for 10
(δέκα),§ and these were repeated as often as necessary.

* Cantor, I., p. 84.   † Cantor, I., p. 113.   ‡ Cantor. I., pp. 113-114.
§ Cantor, I., p. 110,

These numerals are described by the Byzantine gram-
marian Herodianus (A. D. 200) and hence are spoken
of as Herodianic numbers.   Shortly after 500 B. C.
two new systems appeared.   One used the 24 letters
of the Ionic alphabet in their natural order for the
numbers from 1 to 24.   The other arranged these
letters apparently at random but actually in an order
fixed arbitrarily; thus, $a=1$, $\beta=2$, ... , $\iota=10$, $\kappa=$
20, ... , $\rho=100$, $\sigma=200$, etc.   Here too there is
no special symbol for the zero.

The Roman numerals* were probably inherited
from the Etruscans.   The noteworthy peculiarities
are the lack of the zero, the subtractive principle
whereby the value of a symbol was diminished by
placing before it one of lower order ($IV=4$, $IX=9$,
$XL=40$, $XC=90$), even in cases where the language
itself did not signify such a subtraction; and finally
the multiplicative effect of a bar over the numerals.
($\overline{XXX}=30\,000$, $\overline{C}=100\,000$).   Also for certain fractions there were special symbols and names.   Accord-
ing to Mommsen the Roman number-symbols I, V,
X represent the finger, the hand, and the double
hand.   Zangemeister proceeds from the standpoint
that *decem* is related to *decussare* which means a
perpendicular or oblique crossing, and argues that
every straight or curved line drawn across the symbol
of a number in the decimal system multiplies that
number by ten.   In fact, there are on monuments

* Cantor, I., p. 486.

representations of 1, 10, and 1000, as well as of 5 and
500, to prove his assertion.*

Of especial interest in elementary arithmetic is the
number-system of the Hindus, because it is to these
Aryans that we undoubtedly owe the valuable position-
system now in use. Their oldest symbols for 1 to 9
were merely abridged number-words, and the use of
letters as figures is said to have been prevalent from
the second century A. D.† The zero is of later origin ;
its introduction is not proven with certainty till after
400 A. D. The writing of numbers was carried on,
chiefly according to the position-system, in various
ways. One plan, which Aryabhatta records, represented the numbers from 1 to 25 by the twenty-five
consonants of the Sanskrit alphabet, and the succeed-
ing tens (30, 40 . . . . 100) by the semi-vowels and
sibilants. A series of vowels and diphthongs formed
multipliers consisting of powers of ten, *ga* meaning
3, *gi* 300, *gu* 30 000, *gau* $3 \cdot 10^{18}$.‡ In this there is no
application of the position-system, although it ap-
pears in two other methods of writing numbers in
use among the arithmeticians of Southern India.
Both of these plans are distinguished by the fact that

---

*Sitzungsberichte der Berliner Akademie vom 10. November 1887. Words-
worth, in his *Fragments and Specimens of Early Latin*, 1874, derives C for
centum, M for *mille*, and L for *quinquaginta* from three letters of the Chal-
cidian alphabet, corresponding to θ, φ, and χ. He says: "The origin of this
notation is, I believe, quite uncertain, or rather purely arbitrary, though, of
course, we observe that the initials of *mille* and *centum* determined the final
shape taken by the signs, which at first were very different in form."

† See *Encyclopædia Britannica*, under "Numerals"

‡ Cantor, I., p. 566.

the same number can be made up in various ways.
Rules of calculation were clothed in simple verse easy
to hold in mind and to recall. For the Hindu mathematicians this was all the more important since they
sought to avoid written calculation as far as possible.
One method of representation consisted in allowing
the alphabet, in groups of 9 symbols, to denote the
numbers from 1 to 9 repeatedly, while certain vowels
represented the zeros. If in the English alphabet according to this method we were to denote the numbers from 1 to 9 by the consonants b, c, . . . s so that
after two countings one finally has $z = 2$, and were to
denote zero by every vowel or combination of vowels,
the number 60502 might be indicated by *siren* or *heron*,
and might be introduced by some other words in the
text. A second method employed type-words and
combined them according to the law of position.
Thus *abdhi* (one of the 4 seas) = 4, *surya* (the sun
with its 12 houses) = 12, *açvin* (the two sons of the
sun) = 2. The combination *abdhisuryaçvinas* denoted
the number 2124.*

Peculiar to the Sanskrit number-language are special words for the multiplication of very large numbers. *Arbuda* signifies 100 millions, *padma* 10 000
millions; from these are derived *maharbuda* = 1000
millions, *mahapadma* = 100 000 millions. Specially-
formed words for large numbers run up to 10^{17} and
even further. This extraordinary extension of the

* Cantor, I., p. 567.

decimal system in Sanskrit resembles a number-game, a mania to grasp the infinitely great. Of this endeavor to bring the infinite into the realm of number-perception and representation, traces are found also among the Babylonians and Greeks. This appearance may find its explanation in mystic-religious conceptions or philosophic speculations.

The ancient Chinese number-symbols are confined to a comparatively few fundamental elements arranged in a perfectly developed decimal system. Here the combination takes place sometimes by multiplication, sometimes by addition. Thus *san* = 3, *che* = 10; *che san* denotes 13, but *san che* 30.* Later, as a result of foreign influence, there arose two new kinds of notation whose figures show some resemblance to the ancient Chinese symbols. Numbers formed from them were not written from above downward but after the Hindu fashion from left to right beginning with the highest order. The one kind comprising the merchants' figures is never printed but is found only in writings of a business character. Ordinarily the ordinal and cardinal numbers are arranged in two lines one above another, with zeros when necessary, in the form of small circles. In this notation

$$|| = 2, \ \mathsf{X} = 4, \ \underline{\perp} = 6, \ + = 10, \ \hbar = 10\ 000, \ \bigcirc = 0,$$

and hence $\hbar \ \mathrm{o} \ \mathrm{o} \ \overset{\text{\tiny{II}}}{+} \underset{}{\overset{\mathsf{X}}{\underline{\perp}}} = 20\ 046.$

* Cantor, I., p. 630.

Among the Arabs, those skilful transmitters of
Oriental and Greek arithmetic to the nations of the
West, the custom of writing out number-words continued till the beginning of the eleventh century.
Yet at a comparatively early period they had already
formed abbreviations of the number-words, the *Divani*
figures.   In the eighth century the Arabs became acquainted with the Hindu number-system and its figures, including zero.   From these figures there arose
among the Western Arabs, who in their whole literature presented a decided contrast to their Eastern re-
latives, the Gubar numerals (dust-numerals) as vari-
ants. These Gubar numerals, almost entirely forgotten
to-day among the Arabs themselves, are the ancestors
of our modern numerals,* which are immediately derived from the apices of the early Middle Ages. These
primitive Western forms used in the abacus-calcula-
tions are found in the West European MSS. of the
eleventh and twelfth centuries and owe much of their
prominence to Gerbert, afterwards Pope Sylvester II.
(consecrated 999 A. D.).

The arithmetic of the Western nations, cultivated
to a considerable extent in the cloister-schools from
the ninth century on, employed besides the abacus the
Roman numerals, and consequently made no use of a
symbol for zero. In Germany up to the year 1500 the
Roman symbols were called German numerals in distinction from the symbols—then seldom employed—

* Hankel, p. 255.

of Arab-Hindu origin, which included a zero (Arabic
as-sifr, Sanskrit *sunya*, the void). The latter were
called ciphers (*Ziffern*). From the fifteenth century on
these Arab-Hindu numerals appear more frequently in
Germany on monuments and in churches, but at that
time they had not become common property.* The
oldest monument with Arabic figures (in Katharein
near Troppau) is said to date from 1007. Monuments
of this kind are found in Pforzheim (1371), and in Ulm
(1388). A frequent and free use of the zero in the
thirteenth century is shown in tables for the calculation of the tides at London and of the duration of
moonlight.† In the year 1471 there appeared in Cologne a work of Petrarch with page-numbers in Hindu
figures at the top. In 1482 the first German arithmetic with similar page-numbering was published in
Bamberg. Besides the ordinary forms of numerals
everywhere used to-day, which appeared exclusively
in an arithmetic of 1489, the following forms for 4, 5,
7 were used in Germany at the time of the struggle
between the Roman and Hindu notations:

$$8 \cdot 9 \cdot \wedge \cdot$$

The derivation of the modern numerals is illustrated
by the examples below which are taken in succession
from the Sanskrit, the apices, the Eastern Arab, the

* Unger, *Die Methodik der praktischen Arithmetik*, 1888, p. 70. Hereafter
referred to as Unger.

† Günther, *Geschichte des mathematischen Unterrichts im deutschen Mittelalter bis zum Jahr 1525*, 1887, p. 175. Hereafter referred to as Günther.

Western Arab Gubar numerals, the numerals of the eleventh, thirteenth, and sixteenth centuries.*

$$\begin{matrix} \mathcal{C} & \sqcup & \mathsf{H} & \mathsf{R} \\ \mathcal{C} & \mathsf{q} & \wedge & 8 \\ \mathsf{F} & \overset{o}{\underset{\beta}{r}} & \vee & \wedge \\ \mathcal{C} & \mathcal{Y} & \mathcal{7} & \mathcal{2} \\ \mathcal{G} & \omega & \mathsf{V} & \wedge \\ 7 & \mathcal{G} & \wedge & 8 \\ \mathcal{Z} & 5 & 7 & 8 \end{matrix}$$

In the sixteenth century the Hindu position-arithmetic and its notation first found complete introduction among all the civilized peoples of the West. By this means was fulfilled one of the indispensable conditions for the development of common arithmetic in the schools and in the service of trade and commerce.

* Cantor, table appended to **Vol. I**, and Hankel, p. 325.

II. ARITHMETIC.

A. GENERAL SURVEY.

THE simplest number-words and elementary counting have always been the common property of the people. Quite otherwise is it, however, with the different methods of calculation which are derived from simple counting, and with their application to complicated problems. As the centuries passed, that part of ordinary arithmetic which to-day every child knows, descended from the closed circle of particular castes or smaller communities to the common people, so as to form an important part of general culture. Among the ancients the education of the youth had to do almost wholly with bodily exercises. Only a riper age sought a higher cultivation through intercourse with priests and philosophers, and this consisted in part in the common knowledge of to-day: people learned to read, to write, to cipher.

At the beginning of the first period in the historic development of common arithmetic stand the Egyptians. To them the Greek writers ascribe the invention of surveying, of astronomy, and of arithmetic. To their literature belongs also the most ancient book on

arithmetic, that of Ahmes, which teaches operations
with whole numbers and fractions.   The Babylonians
employed a sexagesimal system in their position-arithmetic, which latter must also have served the purposes of a religious number-symbolism. The common
arithmetic of the Greeks, particularly in most ancient
times, was moderate in extent until by the activity of
the scholars of philosophy there was developed a real
mathematical science of predominantly geometric
character.   In spite of this, skill in calculation was
not esteemed lightly.   Of this we have evidence when
Plato demands for his ideal state that the youth should
be instructed in reading, writing, and arithmetic.

The arithmetic of the Romans had a purely practical turn ; to it belonged a mass of quite complicated
problems arising from controversies regarding questions of inheritance, of private property and of reim-
bursement of interest. The Romans used duodecimal
fractions.   Concerning the most ancient arithmetic of
the Hindus only conjectures can be made ; on the contrary, the Hindu elementary arithmetic after the introduction of the position-system is known with toler-
able accuracy from the works of native authors.   The
Hindu mathematicians laid the foundations for the
ordinary arithmetic processes of to-day.   The influence of their learning is perceptible in the Chinese
arithmetic which likewise depends on the decimal system ; in still greater measure, however, among the

Arabs who besides the Hindu numeral-reckoning also
employed a calculation by columns.

The time from the eighth to the beginning of the
fifteenth century forms the second period. This is a
noteworthy period of transition, an epoch of the trans-
planting of old methods into new and fruitful soil,
but also one of combat between the well-tried Hindu
methods and the clumsy and detailed arithmetic operations handed down from the Middle Ages. At
first only in cloisters and cloister-schools could any
arithmetic knowledge be found, and that derived from
Roman sources. But finally there came new suggestions from the Arabs, so that from the eleventh to the
thirteenth centuries there was opposed to the group of
abacists, with their singular complementary methods,
a school of algorists as partisans of the Hindu arithmetic.

Not until the fifteenth century, the period of investigation of the original Greek writings, of the
rapid development of astronomy, of the rise of the
arts and of commercial relations, does the third period in the history of arithmetic begin. As early
as the thirteenth century besides the cathedral and
cloister-schools which provided for their own religious
and ecclesiastical wants, there were, properly speaking, schools for arithmetic. Their foundation is to be
ascribed to the needs of the brisk trade of German
towns with Italian merchants who were likewise skilled
computers. In the fifteenth and sixteenth centuries

school affairs were essentially advanced by the human-
istic tendency and by the reformation. Latin schools,
writing schools, German schools (in Germany) for boys
and even for girls, were established. In the Latin
schools only the upper classes received instruction in
arithmetic, in a weekly exercise : they studied the four
fundamental rules, the theory of fractions, and at most
the rule of three, which may not seem so very little
when we consider that frequently in the universities
of that time arithmetic was not carried much further.
In the writing schools and German boys' schools the
pupils learned something of calculation, numeration,
and notation, especially the difference between the
German numerals (in Roman writing) and the ciphers
(after the Hindu fashion). In the girls' schools, which
were intended only for the higher classes of people, no
arithmetic was taught. Considerable attainments in
computation could be secured only in the schools for
arithmetic. The most celebrated of these institutions
was located at Nuremberg. In the commercial towns
there were accountants' guilds which provided for the
extension of arithmetic knowledge. But real mathematicians and astronomers also labored together in developing the methods of arithmetic. In spite of this
assistance from men of prominence, no theory of arithmetic instruction had been established even as late as
in the sixteenth century. What had been done before had to be copied. In the books on arithmetic

were found only rules and examples, almost never proofs or deductions.

The seventeenth century brought no essential change in these conditions. Schools existed as before where they had not been swallowed up by the horrors of the Thirty-Years' War. The arithmeticians wrote their books on arithmetic, perhaps contrived calculating machines to make the work easier for their pupils, or composed arithmetic conversations and poems. A specimen of this is given in the following extracts from Tobias Beutel's *Arithmetica*, the seventh edition of which appeared in 1693.[*]

> " Numerieren lehrt im Rechen
> Zahlen schreiben und aussprechen."

> "In Summen bringen heisst addieren
> Dies muss das Wörtlein Und vollführen."

> " Wie eine Hand an uns die andre wäschet rein
> Kann eine Species der andern Probe seyn."

> " We are taught in numeration
> Number writing and expression,"
> etc., etc.

Commercial arithmetic was improved by the cultiva tion of the study of exchange and discount, and the abbreviated method of multiplication. The form of instruction remained the same, i. e., the pupil reckoned according to rules without any attempt being made to explain their nature.

The eighteenth century brought as its first and

[*] Unger, p. 124.

most important innovation the statutory regulation of
school matters by special school laws, and the establishment of normal schools (the first in 1732 at Stettin in connection with the orphan asylum). As reor-
ganizers of the higher schools appeared the pietists
and philanthropinists. The former established *Real-
schulen* (the oldest founded 1738 in Halle) and higher
Bürgerschulen; the latter in their *Schulen der Aufklärung*
sought by an improvement of methods to educate
cultured men of the world. The arithmetic exercise-
books of this period contain a simplification of division (the downwards or under-itself division) as well
as a more fruitful application of the chain rule and
decimal fractions. By their side also appear manuals
of method whose number is rapidly increasing in the
nineteenth century. In these, elementary teaching
receives especial attention. According to Pestalozzi
(1803) the foundation of calculation is sense perception, according to Grube (1842), the comprehensive
treatment of each number before taking up the next,
according to Tanck and Knilling (1884), counting.
In Pestalozzi's method "the decimal structure of our
number-system, which includes so many advantages
in the way of calculation, is not touched upon at all,
addition, subtraction, and division do not appear as
separate processes, the accompanying explanations
smother the principal matter in the propositions, that
is the arithmetic truth." * Grube has simply drawn

* Unger, p. 179.

from Pestalozzi's principles the most extreme conclu
sions. His sequence "is in many respects faulty; his
processes unsuitable."* The historical development
of arithmetic speaks in favor of the counting-prin-
ciple : the first reckoning in every age has been an
observing and counting.

B. FIRST PERIOD.

THE ARITHMETIC OF THE OLDEST NATIONS TO THE TIME OF THE ARABS.

1. *The Arithmetic of Whole Numbers.*

If we leave out of account finger-reckoning, which
cannot be shown with absolute certainty, then according to a statement of Herodotus the ancient Egyptian
computation consisted of an operating with pebbles on
a reckoning-board whose lines were at right angles to
the computer. Possibly the Babylonians also used a
similar device. In the ordinary arithmetic of the latter,
as among the Egyptians, the decimal system prevails,
but by its side we also find, especially in dealing with
fractions, a sexagesimal system. This arose without
doubt in the working out of the astronomical observations of the Babylonian priests.† The length of the
year of 360 days furnished the occasion for the division of the circle into 360 equal parts, one of which
was to represent the apparent daily path of the sun
upon the celestial sphere. If in addition the construc-

*Unger, pp. 192, 193.   †Cantor, I., p. 80.

tion of the regular hexagon was known, then it was natural to take every 60 of these parts again as units. The number 60 was called *soss*. Numbers of the sexagesimal system were again multiplied in accordance with the rules of the decimal system : thus a *ner* $= 600$, a *sar* $= 3600$. The sexagesimal system established by the Babylonian priests also entered into their religious speculations, where each of their divinities was designated by one of the numbers from 1 to 60 corresponding to his rank. Perhaps the Babylonians also divided their days into 60 equal parts as has been shown for the Veda calendars of the ancient Hindus.

The Greek elementary mathematics, at any rate as early as the time of Aristophanes (420 B. C.),[*] used finger-reckoning and reckoning-boards for ordinary computation. An explanation of the finger-reckoning is given by Nicholas Rhabda[†] of Smyrna (in the fourteenth century). Moving from the little finger of the left hand to the little finger of the right, three fingers were used to represent units, the next two, tens, the next two, hundreds, and the last three, thousands. On the reckoning board, the *abax* (ᾰβαξ, dust board), whose columns were at right angles to the user, the operations were carried on with pebbles which had a different place-value in different lines. Multiplication was performed by beginning with the highest order in each factor and forming the sum of the partial pro-

---

[*] Cantor, I., pp. 120, 479. [†] Gow, *History of Greek Mathematics*, p. 24.

ducts.   Thus the calculation was effected (in modern form) as follows:

$$126 \cdot 237 = (100 + 20 + 6)(200 + 30 + 7)$$
$$=\quad 20\,000 +\quad 3000\quad + 700$$
$$+\quad 4\,000 +\quad 600\quad + 140$$
$$+\quad 1\,200 +\quad 180\quad +\ 42$$
$$=\quad 29\,862$$

According to Pliny, the finger-reckoning of the Romans goes back to King Numa;* the latter had made a statue of Janus whose fingers represented the number of the days of a year (355). Consistently with this Boethius calls the numbers from 1 to 9 finger-numbers, 10, 20, 30, . . . joint-numbers, 11, 12, . . . 19, 21, 22, . . . 29, . . . composite numbers. In ele

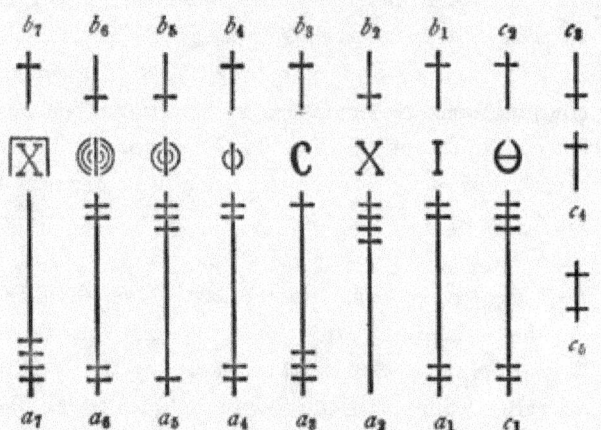

mentary teaching the Romans used the *abacus*, a board usually covered with dust on which one could

*Cantor, I., p. 491.

trace figures, draw columns, and work with pebbles. Or if the abacus was to be used for computing only, it was made of metal and provided with grooves (the vertical lines in the schematic drawing on the preceding page) in which arbitrary marks (the crosslines) could be shifted.

The columns $a_1 \ldots a_7$, $b_1 \ldots b_7$ form a system from 1 to 1 000 000; upon a column a are found four marks, upon a column b only one mark. Each of the four marks represents a unit, but the upper single mark five units of the order under consideration. Further a mark upon $c_1 = \frac{1}{12}$, upon $c_2 = \frac{3}{12}$, upon $c_3 = \frac{1}{24}$, upon $c_4 = \frac{1}{48}$, upon $c_5 = \frac{1}{72}$ (relative to the division of the a's). The abacus of the figure represents the number $782\,192 + \frac{3}{12} + \frac{1}{24} + \frac{1}{72} = 782\,192\frac{11}{36}$. This abacus served for the reckoning of results of simple problems. Along with this the multiplication-table was also employed. For larger multiplications there were special tables. Such a one is mentioned by Victorius (about 450 A. D.).[*] From Boethius, who calls the abacus marks *apices*, we learn something about multiplication and division. Of these operations the former probably, the latter certainly, was performed by the use of complements. In Boethius the term *differentia* is applied to the complement of the divisor to the next complete ten or hundred. Thus for the divisors 7, 84, 213 the differentiæ are 3, 6, 87[†] respectively. The essential characteristics of this comple-

mentary division are seen from the following example
put in modern form :

$$\frac{257}{14} = \frac{257}{20-6} = 10 + \frac{60+57}{20-6} = 10 + \frac{117}{20-6}$$

$$\frac{117}{20-6} = 5 + \frac{30+17}{20-6} = 5 + \frac{47}{20-6}$$

$$\frac{47}{20-6} = 2 + \frac{12+7}{20-6} = 2 + \frac{19}{20-6}$$

$$\frac{19}{14} = 1 + \frac{5}{14}$$

$$\frac{257}{14} = 18 + \frac{5}{14}.$$

The *swanpan* of the Chinese somewhat resembles
the abacus of the Romans. This calculating machine
consists of a frame ordinarily with ten wires inserted.
A cross wire separates each of the ten wires into two
unequal parts; on each smaller part two and on each
larger five balls are strung. The Chinese arithmetics
give no rules for addition and subtraction, but do for
multiplication, which, as with the Greeks, begins
with the highest order, and for division, which appears
in the form of a repeated subtraction.

The calculation of the Hindus, after the introduction of the arithmetic of position, possessed a series
of suitable rules for performing the fundamental operations. In the case of a smaller figure in the minuend subtraction is performed by borrowing and by
addition (as in the so-called Austrian subtraction).*

* The Austrian subtraction corresponds in part to the usual method of
" making change."

In multiplication, for which several processes are available, the product is obtained in some cases by separating the multipliers into factors and subsequently adding the partial products. In other cases a schematic process is introduced whose peculiarities are shown in the example $315 \cdot 37 = 11\ 655$.

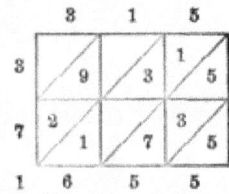

The result of the multiplication is obtained by the addition of the figures found within the rectangle in the direction of the oblique lines. With regard to division we have only a few notices. Probably, however, complementary methods were not used.

The earliest writer giving us information on the arithmetic of the Arabs is Al Khowarazmi. The borrowing from Hindu arithmetic stands out very clearly. Six operations were taught. Addition and subtraction begin with the units of highest order, therefore on the left; halving begins on the right, doubling again on the left. Multiplication is effected by the process which the Hindus called *Tatstha* (it remains standing).* The partial products, beginning with the highest order in the multiplicand, are written above the corresponding figures of the latter and each figure

* Cantor, I., p. 674, 571.

of the product to which other units from a later partial product are added (in sand or dust), rubbed out and corrected, so that at the end of the computation the result stands above the multiplicand. In division, which is never performed in the complementary fashion, the divisor stands below the dividend and advances toward the right as the calculation goes on. Quotient and remainder appear above the divisor in $\frac{461}{16} = 28\frac{13}{16}$, somewhat as follows: [*]

$$
\begin{array}{c}
13 \\
14 \\
28 \\
461 \\
16 \\
16
\end{array}
$$

Al Nasawi[†] also computes after the same fashion as Al Khowarazmi. Their methods characterise the elementary arithmetic of the Eastern Arabs.

In essentially the same manner, but with more or less deviation in the actual work, the Western Arabs computed. Besides the Hindu figure-computation Ibn al Banna teaches a sort of reckoning by columns.[‡] Proceeding from right to left, the columns are combined in groups of three; such a group is called *takarrur*; the number of all the columns necessary to record a number is the *mukarrar*. Thus for the number 3 849 922 the *takarrur* or number of complete groups is 2, the *mukarrar* = 7. Al Kalsadi wrote a

* Cantor, I., p. 674.   † Cantor, I., p. 716.   ‡ Cantor, I., p. 757.

work *Raising of the Veil of the Science of Gubar.*[*] The original meaning of Gubar (dust) has here passed over into that of the written calculation with figures. Especially characteristic is it that in addition, subtraction (= *tarh, taraha* = to throw away) and multiplication the results are written above the numbers operated upon, as in the following examples:

$193 + 45 = 238$ and $238 - 193 = 45$
is written,   is written,

$$\frac{238}{193};\qquad \frac{45}{238}.$$

$$45\qquad\qquad 193$$
$$1\qquad\qquad\quad 1$$

Several rules for multiplication are found in Al Kalsadi, among them one with an advancing multiplier. In division the result stands below.

FIRST EXAMPLE.

$7 \cdot 143 = 1001$
is written, $\dfrac{1001}{21}$

$$28$$
$$\dfrac{7}{143}$$
$$777$$

SECOND EXAMPLE.

$$\frac{1001}{7} = 143$$

is written, 32

$$1001$$
$$\dfrac{777}{143}$$

2. *Calculation With Fractions.*

In his arithmetic Ahmes gives a large number of examples which show how the Egyptians dealt with fractions. They made exclusive use of unit-fractions,

* Cantor, I., p. 762.

i. e., fractions with numerator 1. For this numerator, therefore, a special symbol is found, in the hiero-glyphic writing ⌒, in the hieratic a point, so that in the latter a unit fraction is represented by its denomi-nator with a point placed above it. Besides these there are found for $\frac{1}{2}$ and $\frac{2}{3}$ the hieroglyphs ▭ and 𝍸;[*] in the hieratic writing there are likewise special symbols corresponding to the fractions $\frac{1}{2}$, $\frac{2}{3}$, $\frac{1}{3}$, and $\frac{1}{4}$. The first problem which Ahmes solves is this, to sep-arate a fraction into unit fractions. E. g., he finds $\frac{2}{5}=\frac{1}{3}+\frac{1}{15}$, $\frac{2}{97}=\frac{1}{56}+\frac{1}{679}+\frac{1}{776}$. This separation, really an indeterminate problem, is not solved by Ahmes in general form, but only for special cases.

The fractions of the Babylonians being entirely in the sexagesimal system, had at the outset a com-mon denominator, and could be dealt with like whole numbers. In the written form only the numerator was given with a special sign attached. The Greeks wrote a fraction so that the numerator came first with a single stroke at the right and above, followed in the same line by the denominator with two strokes, writ-ten twice, thus $ιζ'κα''κα''=\frac{17}{21}$. In unit fractions the numerator was omitted and the denominator written only once: $δ''=\frac{1}{4}$. The unit fractions to be added follow immediately one after another.[†] $ζ''κη''ριβ''σκδ''$ $=\frac{1}{7}+\frac{1}{28}+\frac{1}{112}+\frac{1}{224}=\frac{43}{224}$. In arithmetic proper, extensive use was made of unit-fractions, later also of

* For carefully drawn symbols see Cantor, I. p. 45.
† Cantor, I., p. 118.

sexagesimal fractions (in the computation of angles).
Of the use of a bar between the terms of a fraction
there is nowhere any mention. Indeed, where such
use appears to occur, it marks only the result of an
addition, but not a division.[*]

The fractional calculations of the Romans furnish
an example of the use of the duodecimal system.
The fractions (*minutiæ*) $\frac{1}{12}$, $\frac{2}{12}$, ... $\frac{11}{12}$ had special
names and symbols. The exclusive use of these duo-
decimal fractions[†] was due to the fact that the *as*,
a mass of copper weighing one pound, was divided
into twelve *unciæ*. The *uncia* had four *sicilici* and
twenty-four *scripuli*. $1 = as$, $\frac{1}{2} = semis$, $\frac{1}{3} = triens$, $\frac{1}{4} =$
quadrans, etc. Besides the twelfths special names
were given to the fractions $\frac{1}{24}$, $\frac{1}{48}$, $\frac{1}{72}$, $\frac{1}{144}$, $\frac{1}{288}$. The
addition and subtraction of such fractions was com-
paratively simple, but their multiplication very detailed. The greatest disadvantage of this system consisted in the fact that all divisions which did not fit
into this duodecimal system could be represented by
minutiæ either with extreme difficulty or only approximately.

In the computations of the Hindus both unit fractions and derived fractions likewise appear. The denominator stands under the numerator but is not separated from it by a bar. The Hindu astronomers
preferred to calculate with sexagesimal fractions. In
the computations of the Arabs Al Khowarazmi gives

[*]Tannery in *Bibl. Math.* 1886.   [†]Hankel, p. 57.

special words for half, third, . . . ninth (expressible
fractions).* All fractions with denominators non-divis-
ible by 2, 3, . . . 9, are called mute fractions; they
were expressed by a circumlocution, e. g., $\frac{2}{17}$ as 2
parts of 17 parts. Al Nasawi writes mixed numbers
in three lines, one under another, at the top the whole
number, below this the numerator, below this the denominator. For astronomical calculations fractions
of the sexagesimal system were used exclusively.

3. Applied Arithmetic.

The practical arithmetic of the ancients included
besides the common cases of daily life, astronomical
and geometrical problems. The latter will be passed
over here because they are mentioned elsewhere. In
Ahmes problems in partnership are developed and
also the sums of some of the simplest series determined. Theon of Alexandria showed how to obtain
approximately the square root of a number of angle
degrees by the use of sexagesimal fractions and the
gnomon. The Romans were concerned principally
with problems of interest and inheritance. The Hindus had already developed the method of false position (*Regula falsi*) and the rule of three, and made
a study of problems of alligation, cistern-filling, and
series, which were still further developed by the Arabs.
Along with the practical arithmetic appear frequent

* Cantor, I., p. 675.

traces of observations on the theory of numbers. The Egyptians knew the test of divisibility of a number by 2. The Pythagoreans distinguished numbers as odd and even, amicable, perfect, redundant and defective.[*] Of two amicable numbers each was equal to the sum of the aliquot parts of the other (220 gives $1 + 2 + 4 + 5 + 10 + 11 + 20 + 22 + 44 + 55 + 110 = 284$ and 284 gives $1 + 2 + 4 + 71 + 142 = 220$). A perfect number was equal to the sum of its aliquot parts ($6 = 1 + 2 + 3$). If the sum of the aliquot parts was greater or less than the number itself, then the latter was called redundant or defective respectively ($8 > 1 + 2 + 4$; $12 < 1 + 2 + 3 + 4 + 6$). Besides this, Euclid starting from his geometric standpoint commenced some fundamental investigations on divisibility, the greatest common measure and the least common multiple. The Hindus were familiar with casting out the nines and with continued fractions, and from them this knowledge went over to the Arabs. However insignificant may be these beginnings in their ancient form, they contain the germ of that vast development in the theory of numbers which the nineteenth century has brought about.

* Cantor, I., p. 156.

C. SECOND PERIOD.

1. The Arithmetic of Whole Numbers.

In the cloister schools, the episcopal schools, and
the private schools of the Merovingian and Carloving-
ian period it was the monks almost exclusively who
gave instruction.   The cloister schools proper were of
only slight importance in the advancement of mathematical knowledge : on the contrary, the episcopal
and private schools, the latter based on Italian methods, seem to have brought very beneficial results.
The first to foreshadow something of the mathematical knowledge of the monks is Isidorus of Seville.
This cloister scholar confined himself to making conjectures regarding the derivation of the Roman numerals, and says nothing at all about the method of
computation of his contemporaries.   The Venerable
Bede likewise published only some extended observations on finger-reckoning.   He shows how to represent numbers by the aid of the fingers, proceeding
from left to right, and thereby assumes a certain acquaintance with finger-reckoning, mentioning as his
predecessors Macrobius and Isidorus.*   This *calculus
digitalis*, appearing in both the East and the West in

* Cantor, I., p. 778.

exactly the same fashion, played an important part in fixing the dates of church feasts by the priests of that time; at least *computus digitalis* and *computus ecclesiasticus* were frequently used in the same sense.*

With regard to the fundamental operations proper Bede does not express himself. Alcuin makes much of number-mysticism and reckons in a very cumbrous manner with the Roman numerals.† Gerbert was the first to give in his *Regula de abaco computi* actual rules, in which he depended upon the arithmetic part of Boethius's work. What he teaches is a pure abacus-reckoning, which was widely spread by reason of his reputation. Gerbert's abacus, of which we have an accurate description by his pupil Bernelinus, was a table which for the drawing of geometric figures was sprinkled with blue sand, but for calculation was divided into thirty columns of which three were reserved for fractional computations. The remaining twenty-seven columns were separated from right to left into groups of three. At the head of each group stood like-wise from right to left S (*singularis*), D (*decem*), C (*centum*). The number-symbols used, the so-called apices, are symbols for 1 to 9, but without zero. In calculating with this abacus the intermediate operations could be rubbed out, so that finally only the result remained; or the operation was made with counters. The fundamental operations were performed principally by the use of complements, and in this respect

* Günther.   † Günther.

division is especially characteristic. The formation
of the quotient $\frac{100}{3} = 33\frac{1}{3}$ will explain this complementary division.

C	D	S
		6
		4
1	9	9
1.	9.	9.
	4.	
	1.	6.
1.		4.
	4	9.
	1.	6.
		4.
	1.	9.
		4.
	1.	3.
		4.
		7.
		1
	1.	4.
	1.	1.
		4.
		1.
		1.
		1.
		1.
	3	3

C	D	S
		6
1	9	9
1.	9.	9.
		1
	3	3

	10 — 4
1 9 9	10
99+40	
139	10
79	7
9+28	
37	3
19	1
13	1
7	1
1	33

In the example given the complete performance of the complementary division stands on the left; the figures to be rubbed
out as the calculation goes on are indicated by a period on the
right. On the right is found the abacus-division without the formation of the difference in the divisor, below it the explanation of
the complementary division in modern notation.

In the tenth and eleventh centuries there appeared a large number of authors belonging chiefly to the clergy who wrote on abacus-reckoning with apices but without the zero and without the Hindu-Arab methods. In the latter the apices were connected with the abacus itself or with the representation of numbers of one figure, while in the running text the Roman numeral symbols stood for numbers of several figures. The contrast between the apices-plan and the Roman is so striking that Oddo, for example, writes: "If one takes 5 times 7, or 7 times 5, he gets XXXV" (the 5 and 7 written in apices).*

At the time of the abacus-reckoning there arose the peculiar custom of representing by special signs certain numbers which do not appear in the Roman system of symbols, and this use continued far into the Middle Ages. Thus, for example, in the town-books of Greifswald 250 is continually represented by CCC^v †

The abacists with their remarkable methods of division completely dominated Western reckoning up to the beginning of the twelfth century. But then a complete revolution was effected. The abacus, the heir of the *computus*, i. e., the old Roman method of calculation and number-writing, was destined to give way to the algorism with its sensible use of zero and its simpler processes of reckoning, but not without a further struggle.‡ People became pupils of the Western Arabs. Among the names of those who extended

*Cantor, I., p. 846.   †Günther, p. 175.   ‡Günther, p. 107.

Arab methods of calculation stands forth especially
pre-eminent that of Gerhard of Cremona, because he
translated into Latin a series of writings of Greek
and Arab authors.* Then was formed the school of
algorists who in contrast to the abacists possessed no
complementary division but did possess the Hindu
place-system with zero. The most lasting material
for the extension of Hindu methods was furnished by
Fibonacci in his *Liber abaci*. This book "has been
the mine from which arithmeticians and algebraists
have drawn their wisdom; on this account it has become in general the foundation of modern science."†
Among other things it contains the four rules for
whole numbers and fractions in detailed form. It is
worthy of especial notice that besides ordinary subtraction with borrowing he teaches subtraction by increasing the next figure of the subtrahend by one,
and that therefore Fibonacci is to be regarded as the
creator of this elegant method.

2. *Arithmetic of Fractions.*

Here, also, after Roman duodecimal fractions had
been exclusively cultivated by the abacists Beda, Gerbert and Bernelinus, Fibonacci laid a new foundation
in his exercises preliminary to division. He showed
how to separate a fraction into unit fractions. Especially advantageous in dealing with small numbers

* Hankel, p. 336.   † Hankel, p. 343.

is his method of determining the common denominator: the greatest denominator is multiplied by each following denominator and the greatest common measure of each pair of factors rejected. (Example : the least common multiple of 24, 18, 15, 9, 8, 5 is $24 \cdot 3 \cdot 5$ $= 360$.)

3. Applied Arithmetic.

The arithmetic of the abacists had for its main purpose the determination of the date of Easter. Besides this are found, apparently written by Alcuin, *Problems for Quickening the Mind* which suggest Roman models. In this department also Leonardo Fibonacci furnishes the most prominent rule (the *regula falsi*), but his problems belong more to the domain of algebra than to that of lower arithmetic.

Investigations in the theory of numbers could hardly be expected from the school of abacists. On the other hand, the algorist Leonardo was familiar with casting out the nines, for which he furnished an independent proof.

D. THIRD PERIOD.

FROM THE FIFTEENTH TO THE NINETEENTH CENTURY.

1. The Arithmetic of Whole Numbers.

While on the whole the fourteenth century had only reproductions to show, a new period of brisk ac-

tivity begins with the fifteenth century, marked by
Peurbach and Regiomontanus in Germany, and by
Luca Pacioli in Italy. As far as the individual processes are concerned, in addition the sum sometimes
stands above the addends, sometimes below; subtraction recognizes "carrying" and "borrowing"; in
multiplication various methods prevail; in division no
settled method is yet developed. The algorism of
Peurbach names the following arithmetic operations:
*Numeratio, additio, subtractio, mediatio, duplatio, multi-
plicatio, divisio, progressio* (arithmetic and geometric
series), besides the extraction of roots which before the
invention of decimal fractions was performed by the
aid of sexagesimal fractions. His upwards-division
still used the arrangement of the advancing divisor;
it was performed in the manner following (on the left
the explanation of the process, on the right Peurbach's
division, where figures to be erased in the course of
the reckoning are indicated by a period to the right
and below): The oral statement would be somewhat
like this: 36 in 84 twice, $2 \cdot 3 = 6$, $8 — 6 = 2$, written
above 8; $2 \cdot 6 = 12$, $24 — 12 = 12$, write above, strike
out 2, etc. The proof of the accuracy of the result is
obtained as in the other operations by casting out the
nines. This method of upwards-division which is not
difficult in oral presentation is still found in arithmetics which appeared shortly before the beginning
of the nineteenth century.

```
      |36
8479|235
6
24
12
12
9
37
18
19
15
49
30
19
```

```
          1.1
      1.3.4.
      2.2.9.9
      8 4 7 9 | 235
      3 6 6 6
```

In the sixteenth century work in arithmetic had entered the Latin schools to a considerable extent; but to the great mass of children of the common people neither school men nor statesmen gave any thought before 1525. The first regulation of any value in this line is the Bavarian *Schuelordnungk de anno 1548* which introduced arithmetic as a required study into the village schools. Aside from an occasional use of finger-reckoning, this computation was either a computation upon lines with counters or a figure-computation. In both cases the work began with practice in numeration in figures. To perform an operation with counters a series of horizontal parallels was drawn upon a suitable base. Reckoned from below upward each counter upon the 1st, 2d, 3d, . . . line represented the value 1, 10, 100, . . ., but between the lines they represented 5, 50, 500, . . . The following figure shows the representation of 41 096½.   In subtraction the minuend, in multiplication the multiplicand was put upon the lines.   Division was treated as repeated subtractions. This line-reckoning was completely lost in the seven-

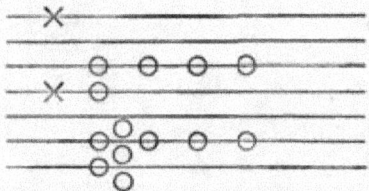

teenth century when it gave place to real written arithmetic or figure-reckoning by which it had been accompanied in the better schools almost from the first.

In the ordinary business and trade of the Middle Ages use was also made of the widely-extended score-reckoning.   At the beginning of the fifteenth century this method was quite usual in Frankfort on the Main, and in England it held its own even into the nine-teenth century.   Whenever goods were bought of a merchant on credit the amount was represented by notches cut upon a stick which was split in two length-wise so that of the two parts which matched, the debtor kept one and the creditor one so that both were se-cured against fraud.*

In the cipher-reckoning the computers of the six-teenth century generally distinguished more than 4 operations; some counted 9, i. e., the 8 named by

*Cantor, M.   *Mathem. Beitr. zum Kulturleben der Völker.*   Halle, 1863.

Peurbach and besides, as a ninth operation, evolution, the extraction of the square root by the formula $(a+b)^2 = a^2 + 2ab + b^2$, and the extraction of the cube root by the formula $(a+b)^3 = a^3 + (a+b)3ab + b^3$. Definitions appeared, but these were only repeated circumlocutions. Thus Grammateus says: "Multiplication shows how to multiply one number by the other. Subtraction explains how to subtract one number from the other so that the remainder shall be seen."[*]

Addition was performed just as is done to-day. In subtraction for the case of a larger figure in the subtrahend, it was the custom in Germany to complete this figure to 10, to add this complement to the minuend figure, but at the same time to increase the figure of next higher order in the subtrahend by 1 (Fibonacci's counting-on method). In more comprehensive books, borrowing for this case was also taught. Multiplication, which presupposed practice in the multiplication table, was performed in a variety of ways. Most frequently it was effected as to-day with a descent in steps by movement toward the left. Luca Pacioli describes eight different kinds of multiplication, among them those above mentioned, with two old Hindu methods, one represented on p. 29, the other cross-multiplication or the lightning method. In the latter method there were grouped all the products involving units, all those involving tens, all those involving hundreds. The multiplication

[*] Unger, p. 72.

$$243 \cdot 139 = 9 \cdot 3 + 10(9 \cdot 4 + 3 \cdot 3) + 100(9 \cdot 2 + 3 \cdot 4 + 1 \cdot 3)$$
$$+ 1000(2 \cdot 3 + 1 \cdot 4) + 10\ 000 \cdot 2 \cdot 1$$

was represented as follows:

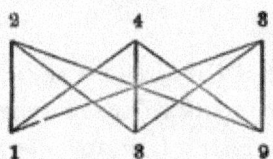

In German books are found, besides these, two noteworthy methods of multiplication, of which one begins on the left (as with the Greeks), the partial products being written in succession in the proper place,
as shown by the following example $243 \cdot 839$:

839
243
——
166867 $839 \cdot 243 = 2 \cdot 8 \cdot 10^4 + 2 \cdot 3 \cdot 10^3 + 2 \cdot 9 \cdot 10^2$
3129 $+ 4 \cdot 8 \cdot 10^3 + 4 \cdot 3 \cdot 10^2 + 4 \cdot 9 \cdot 10$
232                     $+ 3 \cdot 8 \cdot 10^2 + 3 \cdot 3 \cdot 10 + 3 \cdot 9.$
14
2
——
203877

In division the upwards-division prevailed; it was
used extensively, although Luca Pacioli in 1494 taught
the downwards-division in modern form.

After the completion of the computation, in con-
formity to historical tradition, a proof was demanded.
At first this was secured by casting out the nines.
On account of the untrustworthiness of this method,
which Pacioli perfectly realised, the performance of

the inverse operation was recommended. In course of time the use of a proof was entirely given up.

Signs of operation properly so called were not yet in use; in the eighteenth century they passed from algebra into elementary arithmetic. Widmann, however, in his arithmetic has the signs + and —, which had probably been in use some time among the merchants, since they appear also in a Vienna MS. of the fifteenth century.* At a later time Wolf has the sign ÷ for minus. In numeration the first use of the word "million" in print is due to Pacioli (*Summa de Arithmetica*, 1494). Among the Italians the word "million" is said originally to have represented a concrete mass, viz., ten tons of gold. Strangely enough, the words "byllion, tryllion, quadrillion, quyllion, sixlion, septyllion, ottyllion, nonyllion," as well as "million," are found as early as 1484 in Chuquet, while the word "miliars" (equal to 1000 millions) is to be traced back to Jean Trenchant of Lyons (1588).†

The seventeenth century was especially inventive in instrumental appliances for the mechanical performance of the fundamental processes of arithmetic. Napier's rods sought to make the learning of the multiplication-table superfluous. These rods were quadrangular prisms which bore on each side the small multiplication-table for one of the numbers 1, 2, ... 9.

---

* Gerhardt, *Geschichte der Mathematik in Deutschland*, 1877. Hereafter referred to as Gerhardt.

† Müller, *Historisch-etymologische Studien über mathematische Terminologie*. Hereafter referred to as Müller.

For extracting square and cube roots rods were used
with the squares and cubes of one-figure numbers inscribed upon them. Real calculating machines which
gave results by the simple turning of a handle, but on
that account must have proved elaborate and expen-
sive, were devised by Pascal, Leibnitz, and Matthäus
Hahn (1778).

A simplification of another kind was effected by
calculating-tables. These were tables for solving
problems, accompanied also by very extended multi-
plication-tables, such as those of Herwart von Hohen-
burg, from which the product of any two numbers
from 1 to 999 could be read immediately.

For the methods of computation of the eighteenth
century the arithmetic writings of the two Sturms,
and of Wolf and Kästner, are of importance. In the
interest of commercial arithmetic the endeavor was
made to abbreviate multiplication and division by
various expedients. Nothing essentially new was
gained, however, unless it be the so-called mental
arithmetic or oral reckoning which in the later decades
of this period appears as an independent branch.

The nineteenth century has brought as a novelty
in elementary arithmetic only the introduction of the
so-called Austrian subtraction (by counting on) and
division, methods for which Fibonacci had paved the
way. The difference $323 - 187 = 136$ is computed
by saying, 7 and 6, 9 and 3, 2 and 1; and $43083 : 185$
is arranged as in the first of the following examples:

		185
43083	232	
	608	
	533	
	163	

1	1679	2737	1
1	621	1058	1
2	184	437	2
2	46	69	1
	0	23	

With sufficient practice this process certainly secured a considerable saving of time, especially in the case of the determination of the greatest common divisor of two or more numbers as shown by the second of the above examples

$$\frac{1679}{2737} = \frac{73}{119}.$$

2. *Arithmetic of Fractions.*

At the beginning of this period reckoning with fractions was regarded as very difficult. The pupil was first taught how to read fractions: "It is to be noticed that every fraction has two figures with a line between. The upper is called the numerator, the lower the denominator. The expression of fractions is then: name first the upper figure, then the lower, with the little word *part* as $\frac{2}{3}$ part" (Grammateus, 1518).* Then came rules for the reduction of fractions to a common denominator, for reduction to lowest terms, for multiplication and division; in the last the fractions were first made to have a common denominator. Still more is found in Tartaglia who knew how to find the least common denominator; in Stifel who performed

* Unger, p. 84.

division by a fraction by the use of its reciprocal, and in the works of other writers.

The way for the introduction of decimal fractions was prepared by the systems of sexagesimal and duodecimal fractions, since by their employment operations with fractions can readily be performed by the corresponding operations with whole numbers. A notation such as has become usual in decimal fractions was already known to Rudolff,[*] who, in the division of integers by powers of 10, cuts off the requisite number of places with a comma. The complete knowledge of decimal fractions originated with Simon Stevin who extended the position-system below unity to any extent desired. Tenths, hundredths, thousandths, ... were called *primes, sekondes, terzes* . . .; 4.628 is written $4_{(0)} 6_{(1)} 2_{(2)} 8_{(3)}$. Joost Bürgi, in his tables of sines, perhaps independently of Stevin, used decimal fractions in the form 0.32 and 3.2. The introduction of the comma as a decimal point is to be assigned to Kepler.[†] In practical arithmetic, aside from logarithmic computations, decimal fractions were used only in computing interest and in reduction-tables. They were brought into ordinary arithmetic at the beginning of the nineteenth century in connection with the introduction of systems of decimal standards.

---

[*] Gerhardt.

[†] The first use of the decimal point is found in the trigonometric tables of Pitiscus, 1612. Cantor, II., p. 555.

3. *Applied Arithmetic.*

During the transition period of the Middle Ages applied arithmetic had absorbed much from the Latin treatises in a superficial and incomplete manner; the fifteenth and sixteenth centuries show evidences of progress in this direction also. Even the Bamberger Arithmetic of 1483 bears an exclusively practical stamp and aims only at facility of computation in mercantile affairs. That method of solution which in the books on arithmetic everywhere occupied the first place was the "regeldetri" (*regula de tri*, rule of three), known also as the "merchant's rule," or "golden rule."* The statement of the rule of three was purely mechanical; so little thought was bestowed upon the accompanying proportion that even master accountants were content to write 4 fl 12 ℔ 20 fl? instead of $4 \text{ fl} : 20 \text{ fl} = 12 \text{ ℔} : x \text{ ℔}$.† There can indeed be found examples of the rule of three with indirect ratios, but with no explanations of any kind whatever. Problems involving the compound rule of three (*regula de quinque*, etc.) were solved merely by successive applications of the simple rule of three. In Tartaglia and Widmann we find equation of payments treated according to the method still in use to-day. Otherwise, Widmann's Arithmetic of 1489 shows great obscurity and lack of scope in rules and nomenclature, so that not infrequently the same matter appears un-

* Cantor, II., p. 205: Unger, p. 86.   † Cantor, II., p. 368; Unger, p. 87.

der different names. He introduces "Regula Residui,
Reciprocationis, Excessus, Divisionis, Quadrata, In-
ventionis, Fusti, Transversa, Ligar, Equalitatis, Legis,
Augmenti, Augmenti et Decrementi, Sententiarum,
Suppositionis, Collectionis, Cubica, Lucri, Pagamenti,
Alligationis, Falsi," so that in later years Stifel did not
hesitate to declare these things simply laughable.*
Problems of proportional parts and alligation were
solved by the use of as many proportions as corre-
sponded to the number of groups to be separated.
For the computation of compound interest Tartaglia
gave four methods, among them computation by steps
from year to year, or computation with the aid of
the formula $b = aq^n$, although he does not give this
formula. Computing of exchange was taught in its
most simple form. It is said that bills of exchange
were first used by the Jews who migrated into Lombardy after being driven from France in the seventh
century. The Ghibellines who fled from Lombardy
introduced exchange into Amsterdam, and from this
city its use spread.† In 1445 letters of exchange were
brought to Nuremberg.

The chain rule (*Kettensatz*), essentially an Indian
method which is described by Brahmagupta, was developed during the sixteenth century, but did not
come into common use until two centuries later. The
methods of notation differed. Pacioli and Tartaglia

---

*Treutlein, *Die deutsche Coss*, Schlömilch's *Zeitschrift*, Bd. 24, Hl. A.
†Unger, p. 90.

wrote all numbers in a horizontal line and multiplied terms of even and of odd order into separate products. Stifel proceeded in the same manner, only he placed all terms vertically beneath one another. In the work of Rudolff, who also saw the advantage of cancellation, we find the modern method of representing the chain rule, but the answer comes at the end.*

About this time a new method of reckoning was introduced from Italy into Germany by the merchants, which came to occupy an important place in the sixteenth century, and still more so in the seventeenth. This Welsh (i. e., foreign) practice, as it soon came to be called, found its application in the development of the product of two terms of a proportion, especially when these were unlike quantities. The multiplier, together with the fraction belonging to it, was separated into its addends, to be derived successively one from another in the simplest possible manner. How well Stifel understood the real significance and applicability of the Welsh practice, the following statement shows:† "The Welsh practice is nothing more than a clever and entertaining discovery in the rule of three. But let him who is not acquainted with the Welsh practice rely upon the simple rule of three, and he will arrive at the same result which another obtains through the Welsh practice." At this time, too, we find tables of prices and tables of interest in use, their introduction being also ascribable to the Italians.

---

*Unger, p. 92.　　　†Unger, p. 94.

In the sixteenth century we also come upon **examples**
for the *regula virginum* and the *regula falsi* in writings
intended for elementary instruction in arithmetic,—
writings into which, ordinarily, was introduced all the
learning of the author. The significance of these
rules, however, does not lie in the realm of elementary arithmetic, but in that of equations. In the
same way, a few arithmetic writings contained directions for the construction of magic squares, and most
of them also contained, as a side-issue, certain arithmetic puzzles and humorous questions (Rudolff calls
them *Schimpfrechnung*). The latter are often mere
disguises of algebraic equations (the problem of the
hound and the hares, of the keg with three taps, of
obtaining a number which has been changed by certain operations, etc.).

The seventeenth century brought essential innovations only in the province of commercial computation.
While the sixteenth century was in possession of correct methods in all computations of interest when
the amount at the end of a given time was sought,
there were usually grave blunders when the principal
was to be obtained, that is, in computing the discount
on a given sum. The discount in 100 was computed
somewhat in this manner:* 100 dollars gives after
two years 10 dollars in interest; if one is to pay the
100 dollars immediately, deduct 10 dollars." No less
a man than Leibnitz pointed out that the discount

* Unger, p. 132.

must be reckoned upon 100. Among the majority of arithmeticians his method met with the misunderstanding that if the discount at 5% for one year is $\frac{1}{21}$, the discount for two years must be $\frac{2}{21}$. It was not until the eighteenth century, after long and sharp controversy, that mathematicians and jurists united upon the correct formula.

In the computation of exchange the Dutch were essentially in advance of other peoples. They possessed special treatises in this line of commercial arithmetic and through them they were well acquainted with the fundamental principles of the arbitration of exchange. In the way of commercial arithmetic many expedients were discovered in the eighteenth century to aid in the performance of the fundamental operations and in solving concrete problems. Calculation of exchange and arbitration of exchange were firmly established and thoroughly discussed by Clausberg. Especial consideration was given to what was called the Reesic rule, which was looked upon as differing from the well-known chain-rule. Rees's book, which was written in Dutch, was translated into French in 1737, and from this language into German in 1739. In the construction of his series Rees began with the required term; in the computation the elimination of fractions and cancellation came first, and then followed the remaining operations, multiplication and division.

Computation of capital and interest was extended,

through the establishment of insurance associations,
to a so-called political arithmetic, in which calculation of contingencies and annuities held an important
place.

The first traces of conditions for the evolution of
a political arithmetic* date back to the Roman prefect
Ulpian, who about the opening of the third century
A.D. projected a mortality table for Roman subjects.†
But there are no traces among the Romans of life insurance institutions proper. It is not until the Middle
Ages that a few traces appear in the legal regulations
of endowments and guild finances. From the fourteenth century there existed travel and accident insurance companies which bound themselves, in consideration of the payment of a certain sum, to ransom
the insured from captivity among the Turks or Moors.

Among the guilds of the Middle Ages the idea of
association for mutual assistance in fires, loss of cattle
and similar losses had already assumed definite shape.
To a still more marked degree was this the case among
the guilds of artisans which arose after the Reformation—guilds which established regular sick and burial
funds.

We must consider tontines as the forerunner of
annuity insurance. In the middle of the seventeenth
century an Italian physician, Lorenzo Tonti, induced
a number of persons in Paris to contribute sums of

---

* Karup, *Theoretisches Handbuch der Lebensversicherung.* 1871.
† Cantor, 1., p. 522.

money the interest of which should be divided annu-
ally among the surviving members. The French gov-
ernment regarded this procedure as an easy method
of obtaining money and established from 1689 to 1759
ten state tontines which, however, were all given up
in 1770, as it had been proved that this kind of state
loan was not lucrative.

In the meantime two steps had been taken which,
by using the results of mathematical science, provided
a secure foundation for the business of insurance.
Pascal and Fermat had outlined the calculation of
contingencies, and the Dutch statesman De Witt had
made use of their methods to lay down in a separate
treatise the principles of annuity insurance based upon
the birth and death lists of several cities of Holland.
On the other hand, Sir William Petty, in 1662, in a
work on political arithmetic* contributed the first valuable investigations concerning general mortality—a
work which induced John Graunt to construct mortality tables. Mortality tables were also published by
Kaspar Neumann, a Breslau clergyman, in 1692, and
these attracted such attention that the Royal Society
of London commissioned the astronomer Halley to
verify these tables. With the aid of Neumann's material Halley constructed the first complete tables of
mortality for the various ages. Although these tables
did not obtain the recognition they merited until half
a century later, they furnished the foundation for all

* Recently republished in inexpensive form in Cassell's National Library.

later works of this kind, and hence Halley is justly called the inventor of mortality tables.

The first modern life-insurance institutions were products of English enterprise. In the years 1698 and 1699 there arose two unimportant companies whose field of operations remained limited. In the year 1705, however, there appeared in London the "Amicable" which continued its corporate existence until 1866. The "Royal Exchange" and "London Assurance Corporation," two older associations for fire and marine insurance, included life insurance in their business in 1721, and are still in existence. There was soon felt among the managers of such institutions the imperative need for reliable mortality tables, a fact which resulted in Halley's work being rescued from oblivion by Thomas Simpson, and in James Dodson's projecting the first table of premiums, on a rising scale, after Halley's method. The oldest company which used as a basis these scientific innovations was the "Society for Equitable Assurances on Lives and Survivorships," founded in 1765.

While at the beginning of the nineteenth century eight life insurance companies were already carrying on their beneficent work in England, there was at the same time not a single institution of this kind upon the Continent, in spite of the progress which had been made in the science of insurance by Leibnitz, the Bernoullis, Euler and others. In France there appeared in 1819 "La compagnie d' assurances générales sur la vie." In Bremen the founding of a life insurance company was frustrated by the disturbances of the war in 1806. It was not until 1828 that the two oldest German companies were formed, the one in Lübeck, the other in Gotha under the management of Ernst Wilhelm Arnoldi, the "Father of German Insurance."

The nineteenth century has substantially enriched the literature of mortality tables, in such tables as those compiled by the Englishmen Arthur Morgan (in the eighteenth century) and Farr, by the Belgian Quetelet, and by the Germans, Brune, Heym, Fischer, Wittstein, and Scheffler. A recent acquisition in this field is the table of deaths compiled in accordance with the vote of the international statistical congress at Budapest in 1876, which gives the mortality of the population of the German Empire for the ten years 1871–1881. Further development and advancement of the science of insurance is provided for by the "Institute of Actuaries" founded in London in 1849— an academic school with examinations in all branches of the subject. There has also been in Berlin since 1868 a "College of the Science of Insurance," but it offers no opportunity for study and no examinations.

The following compilations furnish a survey of the conditions of insurance in the year 1890 and of its development in Germany.* There were in Germany:

---

* Karup, *Theoretisches Handbuch der Lebensversicherung*, 1871. Johnson, *Universal Cyclopædia*, under "Life-Insurance."

AT THE BEGINNING OF THE YEAR	NUMBER OF LIFE INS. CO'S.	NUMBER OF PERSONS INSURED	FOR THE SUM IN ROUND NUMBERS (MILLION MARKS)
1852	12	46,980	170
1858	20	90,128	300
1866	32	305,433	900
1890	49	....	4250

There were in 1890:

IN	NUMBER OF LIFE INSURANCE CO'S.	AMOUNT OF INSURANCE IN FORCE	
Germany......................	49	4250	million marks
Great Britain and Ireland......	75	900	" pounds
France........................	17	3250	" francs
Rest of Europe................	58	3200	" francs
United States of America......	48	4000	" dollars

All that the eighteenth century developed or discovered has been further advanced in the nineteenth. The center of gravity of practical calculation lies in commercial arithmetic. This is also finding expression in an exceedingly rich literature which has been extended in an exhaustive manner in all its details, but which contains nothing essentially new except the methods of calculating interest in accounts current.

III. ALGEBRA.

A. GENERAL SURVEY.

THE beginnings of general mathematical science are the first important outcome of special studies of number and magnitude; they can be traced back to the earliest times, and their circle has only gradually been expanded and completed. The first period reaches up to and includes the learning of the Arabs; its contributions culminate in the complete solution of the quadratic equation of one unknown quantity, and in the trial method, chiefly by means of geometry, of solving equations of the third and fourth degrees.

The second period includes the beginning of the development of the mathematical sciences among the peoples of the West from the eighth century to the middle of the seventeenth. The time of Gerbert forms the beginning and the time of Kepler the end of this period. Calculations with abstract quantities receive a material simplification in form through the use of abbreviated expressions for the development of formulæ; the most important achievement lies in the purely algebraic solution of equations of the third and fourth degrees by means of radicals.

The third period begins with Leibnitz and Newton and extends from the middle of the seventeenth century to the present time. In the first and larger part of this period a new light was diffused over fields which up to that time had been only partially explored, by the discovery of the methods of higher analysis. At the end of this first epoch there appeared certain mathematicians who devoted themselves to the study of combinations but failed to reach the lofty points of view of a Leibnitz. Euler and Lagrange, thereupon, assumed the leadership in the field of pure analysis. Euler led the way with more than seven hundred dissertations treating all branches of mathematics. The name of the great Gauss, who drew from the works of Newton and Euler the first nourishment for his creative genius, adorns the beginning of the second epoch of the third period. Through the publication of more than fifty large memoirs and a number of smaller ones, not alone on mathematical subjects but also on physics and astronomy, he set in motion a multitude of impulses in the most varied directions. At this time, too, there opened new fields in which men like Abel, Jacobi, Cauchy, Dirichlet, Riemann, Weierstrass and others have made a series of most beautiful discoveries.

B. FIRST PERIOD.

FROM THE EARLIEST TIMES TO THE ARABS.

1. General Arithmetic.

However meagre the information which describes
the evolution of mathematical knowledge among the
earliest peoples, still we find isolated attempts among
the Egyptians to express the fundamental processes
by means of signs.   In the earliest mathematical papyrus * we find as the sign of addition a pair of walk-
ing legs travelling in the direction toward which the
birds pictured are looking.   The sign for subtraction
consists of three parallel horizontal arrows.   The sign
for equality is ⫽.   Computations are also to be found
which show that the Egyptians were able to solve simple problems in the field of arithmetic and geometric
progressions.   The last remark is true also of the
Babylonians.   They assumed that during the first five
of the fifteen days between new moon and full moon,
the gain in the lighted portion of its disc (which was
divided into 240 parts) could be represented by a geo
metric progression, during the ten following days by
an arithmetic progression.   Of the 240 parts there
were visible on the first, second, third . . . fifteenth
day

* Cantor, I., p. 37.

5   10   20   40 1.20
1.36 1.52 2.08 2.24 2.40
2.56 3.12 3.28 3.44 4.

The system of notation is sexagesimal, so that we are
to take $3.28 = 3 \times 60 + 28 = 208$.* Besides this there
have been found on ancient Babylonian monuments
the first sixty squares and the first thirty-two cubes
in the sexagesimal system of notation.

The spoils of Greek treasures are far richer. Even
the name of the entire science ἡ μαθηματική comes from
the Greek language. In the time of Plato the word
μαθήματα included all that was considered worthy of
scientific instruction. It was not until the time of the
Peripatetics, when the art of computation (*logistic*)
and arithmetic, plane and solid geometry, astronomy
and music were enumerated in the list of mathematical sciences, that the word received its special significance. Especially with Heron of Alexandria logistic
appears as elementary arithmetic, while arithmetic so
called is a science involving the theory of numbers.

Greek arithmetic and algebra appeared almost
always under the guise of geometry, although the
purely arithmetic and algebraic method of thinking
was not altogether lacking, especially in later times.
Aristotle† is familiar with the representation of quantities by letters of the alphabet, even when those
quantities do not represent line-segments ; he says in

---

*Cantor, I., p. 81.   †Cantor, I., p. 240.

one place: "If A is the moving force, B that which is moved, Γ the distance, and Δ the time, etc." By the time of Pappus there had already been developed a kind of reckoning with capital letters, since he was able to distinguish as many general quantities as there were such letters in the alphabet. (The small letters α, β, γ, stood for the numbers 1, 2, 3, . . .) Aristotle has a special word for "continuous" and a definition for continuous quantities. Diophantus went farther than any of the other Greek writers. With him there already appear expressions for known and unknown quantities. Hippocrates calls the square of a number δύναμις (power), a word which was transferred to the Latin as *potentia* and obtained later its special mathematical significance. Diophantus gives particular names to all powers of unknown quantities up to the sixth, and introduces them in abbreviated forms, so that x^2, x^3, x^4, x^5, x^6, appear as δ͡ϋ, κ͡ϋ, δδ͡ϋ, δκ͡ϋ, κκ͡ϋ. The sign for known numbers is μ͡ϋ. In subtraction Diophantus makes use of the sign ⋔ (an inverted and abridged ψ); ɩ, an abbreviation for ἴσοι, equal, appears as the sign of equality. A term of an expression is called εἶδος; this word went into Latin as *species* and was used in forming the title *arithmetica speciosa* = algebra.[*] The formulæ are usually given in words and are represented geometrically, as long as they have to do only with expressions of the second dimension. The first ten propositions in the second book of Euclid,

* Cantor, I., p. 442.

for example, are enunciations in words and geometric
figures, and correspond among others to the expressions $a(b+c+d...)=ab+ac+ad+.....$, $(a+b)^2$
$=a^2+2ab+b^2=(a+b)a+(a+b)b$.

Geometry was with the Greeks also a means for investigations in the theory of numbers. This is seen,
for instance, in the remarks concerning gnomon-num-
bers. Among the Pythagoreans a square out of which
a corner was cut in the shape of a square was called a
gnomon. Euclid also used this expression for the
figure $ABCDEF$ which is obtained from the parallelogram $ABCB'$ by cutting out the parallelogram $DB'FE$.
The gnomon-number of the Pythagoreans is $2n+1$;
for when $ABCB'$ is a square, the square upon $DE=n$

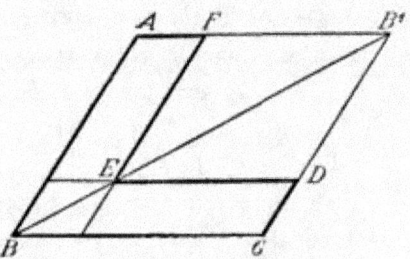

can be made equal to the square on $BC=n+1$ by
adding the square $BE=1\times1$ and the rectangles AE
$=CE=1\times n$, since we have $n^2+2n+1=(n+1)^2$.
Expressions like plane and solid numbers used for
the contents of spatial magnitudes of two and three
dimensions also serve to indicate the constant tend-

ency to objectify mathematical thought by means of geometry.

All that was known concerning numbers up to the third century B. C., Euclid comprehended in a general survey. In his *Elements* he speaks of magnitudes, without, however, explaining this concept, and he understands by this term, besides lines, angles, surfaces and solids, the natural numbers.[*] The difference between even and odd, between prime and composite numbers, the method for finding the least common multiple and the greatest common divisor, the construction of rational right angled triangles according to Plato and the Pythagoreans—all these are familiar to him. A method (the "sieve") for sorting out prime numbers originated with Eratosthenes. It consists in writing down all the odd numbers from 3 on, and then striking out all multiples of 3, 5, 7 . . . Diophantus stated that numbers of the form $a^2 + 2ab + b^2$ represent a square and also that numbers of the form $(a^2 + b^2)(c^2 + d^2)$ can represent a sum of two squares in two ways; for $(ac + bd)^2 + (ad - bc)^2 = (ac - bd)^2 + (ad + bc)^2 = (a^2 + b^2)(c^2 + d^2)$.

The knowledge of the Greeks in the field of elementary series was quite comprehensive. The Pythagoreans began with the series of even and odd numbers. The sum of the natural numbers gives the triangular number, the sum of the odd numbers the square, the sum of the even numbers gives the hetero-

[*] Treutlein.

mecic (oblong) number of the form $n(n+1)$. Square
numbers they also recognised as the sum of two successive triangular numbers. The Neo-Pythagoreans
and the Neo-Platonists made a study not only of polygonal but also of pyramidal numbers. Euclid treated
geometrical progressions in his *Elements*. He obtained the sum of the series $1+2+4+8\ldots$ and
noticed that when the sum of this series is a prime
number, a "perfect number" results from multiplying it by the last term of the series ($1+2+4=7$;
$7\times4=28$; $28=1+2+4+7+14$; cf. p. 35). Infinite convergent series appear frequently in the works
of Archimedes in the form of geometric series whose
ratios are proper fractions; for example, in calculating
the area of the segment of a parabola, where the value
of the series $1+\frac{1}{4}+\frac{1}{16}+\ldots$ is found to be $\frac{4}{3}$. He
also performs a number of calculations for obtaining
the sum of an infinite series for the purpose of estimating areas and volumes. His methods are a substitute for the modern methods of integration, which
are used in cases of this kind, so that expressions like

$$\int_0^c x\,dx=\tfrac{1}{2}c^2, \quad \int_0^c x^2\,dx=\tfrac{1}{3}c^3$$

and other similar expressions are in their import and
essence quite familiar to him.[*]

The introduction of the irrational is to be traced
back to Pythagoras, since he recognised that the hy-

[*] Zeuthen, *Die Lehre von den Kegelschnitten im Altertum.* Deutsch von
v. Fischer-Benzon, 1886.

potenuse of a right-angled isosceles triangle is incommensurable with its sides. The Pythagorean Theodorus of Cyrene proved the irrationality of the square roots of 3, 5, 7, . . . 17.*

Archytas classified numbers in general as rational and irrational. Euclid devoted to irrational quantities a particularly exhaustive investigation in his *Elements*, a work which belongs to the domain of Arithmetic as much as to that of Geometry. Three books among the thirteen, the seventh, eighth and ninth, are of purely arithmetic contents, and in the tenth book there appears a carefully wrought-out theory of "Incommensurable Quantities," that is, of irrational quantities, as well as a consideration of geometric ratios. At the end of this book Euclid shows in a very ingenious manner that the side of a square and its diagonal are incommensurable; the demonstration culminates in the assertion that in the case of a rational relationship between these two quantities a number must have at the same time the properties of an even and an odd number.† In his measurement of the circle Archimedes calculated quite a number of approximate values for square roots; for example,

$$\frac{1351}{780} > \sqrt{3} > \frac{265}{153}.$$

Nothing definite, however, is known concerning the

* Cantor, I., p. 170.

† Montucla, I., p. 208. Montucla says that he knew an architect who lived in the firm conviction that the square root of 2 could be represented as a ratio of finite integers, and who assured him that by this method he had already reached the 100th decimal.

method he used.   Heron also was acquainted with such approximate values ($\frac{7}{5}$ instead of $\sqrt{2}$, $\frac{26}{15}$ instead of $\sqrt{3}$);* and although he did not shrink from the labor of obtaining approximate values for square roots, in the majority of cases he contented himself with the well-known approximation $\sqrt{a^2 \pm b} = a \pm \dfrac{b}{2a}$, e. g., $\sqrt{63} = \sqrt{8^2-1} = 8 - \frac{1}{16}$.   In case greater exactness was necessary, Heron† used the formula $\sqrt{a^2+b} = a + \frac{1}{x} + \frac{1}{y} + \frac{1}{z} + \ldots$ Incidentally he used the identity $\sqrt{a^2 b} = a\sqrt{b}$ and asserted, for example, that $\sqrt{108} = \sqrt{6^2 \cdot 3} = 6\sqrt{3} = 6 \cdot \frac{26}{15} = 10 + \frac{1}{2} + \frac{1}{15}$. Moreover, we find in Heron's *Stereometrica* the first example of the square root of a negative number, namely $\sqrt{81-144}$, which, however, without further consideration, is put down by the computer as 8 less $\frac{1}{16}$, which shows that negative quantities were unknown among the Greeks.   It is true that Diophantus employed differences, but only those in which the minuend was greater than the subtrahend.   Through Theon we are made acquainted with another method of extracting the square root; it corresponds with the method in use at present, with the exception that the Babylonian sexagesimal fractions are used, as was customary until the introduction of decimal fractions.

Furthermore, we find in Aristotle traces of the theory of combinations, and in Archimedes an attempt at the representation of a quantity which in-

---

*Cantor, I., p. 368.   † Tannery in *Bordeaux Mém.*, IV., 1881.

creases beyond all limits, first in his extension of the
number-system, and then in his work entitled ψαμ-
μίτης (Latin *arenarius*, the sand-reckoner). Archimedes arranges the first eight orders of the decimal
system together in an octad; 10^8 octads constitute a
period, and then these periods are arranged again
according to the same law. In the sand-reckoning,
Archimedes solves the problem of estimating the
number of grains of sand that can be contained in a
sphere which includes the whole universe. He assumes that 10,000 grains of sand take up the space of
a poppy-seed, and he finds the sum of all the grains
to be 10 000 000 units of the eighth period of his system, or 10^{63}. It is possible that Archimedes in these
observations intended to create a counterpart to the
domain of infinitesimal quantities which appeared in
his summations of series, a counterpart not accessible
to the ordinary arithmetic.

In the fragments with which we are acquainted
from the writings of Roman surveyors (*agrimensores*)
there are but few arithmetic portions, these having
to do with polygonal and pyramidal numbers. Obviously they are of Greek origin, and the faulty style
in parts proves that there was among the Romans no
adequate comprehension of matters of this kind.

The writings of the Hindu mathematicians are exceedingly rich in matters of arithmetic. Their symbolism was quite highly developed at an early date.*

* Cantor, I., p. 558.

Aryabhatta calls the unknown quantity *gulika* ("little ball"), later *yavattavat*, or abbreviated *ya* ("as much as"). The known quantity is called *rupaka* or *ru* ("coin"). If one quantity is to be added to another, it is placed after it without any particular sign. The same method is followed in subtraction, only in this case a dot is placed over the coefficient of the subtrahend so that positive (*dhana*, assets) and negative quantities (*kshaya*, liabilities) can be distinguished. The powers of a quantity also receive special designations. The second power is *varga* or *va*, the third *ghana* or *gha*, the fourth *va va*, the fifth *va gha ghata*, the sixth *va gha*, the seventh *va va gha ghata* (*ghata* signifies addition). The irrational square root is called *karana* or *ka*. In the Çulvasutras, which are classed among the religious books of the Hindus, but which in addition contain certain arithmetic and geometric deductions, the word *karana* appears in conjunction with numerals; *dvikarani* = $\sqrt{2}$, *trikarani* = $\sqrt{3}$, *daçakarani* = $\sqrt{10}$. If several unknown quantities are to be distinguished, the first is called *ya;* the others are named after the colors: *kalaka* or *ka* (black), *nilaka* or *ni* (blue), *pitaka* or *pi* (yellow); for example, by *ya kabha* is meant the quantity $x \cdot y$, since *bhavita* or *bha* indicates multiplication. There is also a word for "equal"; but as a rule it is not used, since the mere placing of a number under another denotes their equality.

In the extension of the domain of numbers to include negative quantities the Hindus were certainly successful. They used them in their calculations, and obtained them as roots of equations, but never regarded them as proper solutions. Bhaskara was even aware that a square root can be both positive and negative, and also that $\sqrt{-a}$ does not exist for the ordinary number-system. He says: "The square of a positive as well as of a negative number is positive, and the square root of a positive number is double, positive, and negative. There can be no square root of a negative number, for this is no square." [*]

The fundamental operations of the Hindus, of which there were six, included raising to powers and extracting roots. In the extraction of square and cube roots Aryabhatta used the formulæ for $(a + b)^2$ and $(a + b)^3$, and he was aware of the advantage of separating the number into periods of two and three figures each, respectively. Aryabhatta called the square root *varga mula*, and the cube root *ghana mula* (*mula*, root, used also of plants). Transformations of expressions involving square roots were also known. Bhaskara applied the formula[†]

$$\sqrt{a + \sqrt{b}} = \sqrt{\tfrac{1}{2}\left(a + \sqrt{a^2 - b}\right)} + \sqrt{\tfrac{1}{2}\left(a - \sqrt{a^2 - b}\right)},$$

and was also able to reduce fractions with square roots in the denominator to forms having a rational denominator. In some cases the approximation methods for square root closely resemble those of the Greeks.

[*] Cantor, I., p. 585.   [†] Cantor, I., p. 586.

Problems in transpositions, of which only a few traces are found among the Greeks, occupy considerable attention among the Indians. Bhaskara made use of formulæ for permutations and combinations* with and without repetitions, and he was acquainted with quite a number of propositions involving the theory of numbers, which have reference to quadratic and cubic remainders as well as to rational right-angled triangles. But it is noticeable that we discover among the Indians nothing concerning perfect, amicable, defective, or redundant numbers. The knowledge of figurate numbers, which certain of the Greek schools cultivated with especial zeal, is likewise wanting. On the contrary, we find in Aryabhatta, Brahmagupta and Bhaskara summations of arithmetic series, as well as of the series $1^2 + 2^2 + 3^2 + \ldots, 1^3 + 2^3 + 3^3 + \ldots$ The geometric series also appears in the works of Bhaskara. As regards calculation with zero, Bhaskara was aware that $\frac{a}{0} = \infty$.

The Chinese also show in their literature some traces of arithmetic investigations; for example, the binomial coefficients for the first eight powers are given by Chu shi kih in the year 1303 as an "old method." There is more to be found among the Arabs. Here we come at the outset upon the name of Al Khowarazmi, whose Algebra, which was probably translated into Latin by Æthelhard of Bath, opens

*Cantor, I., p. 579.

with the words* "Al Khowarazmi has spoken." In the Latin translation this name appears as *Algoritmi*, and to-day appears as *algorism* or *algorithm*, a word completely separated from all remembrance of Al Khowarazmi, and much used for any method of computation commonly employed and proceeding according to definite rules. In the beginning of the sixteenth century there appears in a published mathematical work a *"philosophus nomine Algorithmus,"* a sufficient proof that the author knew the real meaning of the word algorism. But after this, all knowledge of the fact seems to disappear, and it was not until our own century that it was rediscovered by Reinaud and Boncompagni.†

Al Khowarazmi increased his knowledge by studying the Greek and Indian models. A known quantity he calls a number, the unknown quantity *jidr* (root) and its square *mal* (power). In Al Karkhi we find the expression *kab* (cube) for the third power, and there are formed from these expressions $mal\ mal = x^4$, $mal\ kab = x^5$, $kab\ kab = x^6$, $mal\ mal\ kab = x^7$, etc. He also treats simple expressions with square roots, but without arriving at the results of the Hindus. There is a passage in Omar Khayyam from which it is to be inferred that the extraction of roots was always performed by the help of the formula for $(a + b)^n$. Al Kalsadi‡ contributed something new by the introduc-

* Cantor, I., p. 671.
† *Jahrbuch über die Fortschritte der Mathematik*, 1887, p. 23.
‡ Cantor, I., p. 765.

tion of a radical sign. Instead of placing the word
jidr before the number of which the square root was
to be extracted, as was the custom, Al Kalsadi makes
use only of the initial letter ح of this word and places
it over the number, as,

$$\overset{\cdot}{2} = \sqrt{2}, \quad \tfrac{1}{2}\overset{\cdot}{2} = \sqrt{2\tfrac{1}{2}}, \quad \overset{2}{5} = 2\sqrt{5}.$$

Among the Eastern Arabs the mathematicians
who investigated the theory of numbers occupied
themselves particularly with the attempt to discover
rational right-angled triangles and with the problem
of finding a square which, if increased or diminished
by a given number, still gives a square. An anony-
mous writer, for example, gave a portion of the theory of quadratic remainders, and Al Khojandi also
demonstrated the proposition that upon the hypothesis of rational numbers the sum of two cubes cannot
be another third power. There was also some knowledge of cubic remainders, as is seen in the application by Avicenna of the proof by excess of nines in
the formation of powers. This mathematician gives
propositions which can be briefly represented in the
form*

$$(9n \pm 1)^2 \equiv 1 \pmod 9), \quad (9n \pm 2)^2 \equiv 4 \pmod 9),$$
$$(9n + 1)^3 \equiv (9n + 4)^3 \equiv (9n + 7)^3 \equiv 1 \pmod 9), \text{ etc.}$$

Ibn al Banna has deductions of a similar kind which
form the basis of a proof by eights and sevens.†

In the domain of series the Arabs were acquainted

*Cantor, I., p. 712.   † Cantor, I., p. 759.

at least with arithmetic and geometric progressions and also with the series of squares and cubes. In this field Greek influence is unmistakable.

2. *Algebra.*

The work of Ahmes shows that the Egyptians were possessed of equations of the first degree, and used in their solution methods systematically chosen. The unknown x is called *hau* (heap); an equation* appears in the following form: *heap*, its $\frac{2}{3}$, its $\frac{1}{2}$, its $\frac{1}{7}$, its whole, gives 37, that is $\frac{2}{3}x + \frac{1}{2}x + \frac{1}{7}x + x = 37$.

The ancient Greeks were acquainted with the solution of equations only in geometrical form. Nowhere, save in proportions, do we find developed examples of equations of the first degree which would show unmistakably that the root of a linear equation with one unknown was ever determined by the intersection of two straight lines; but in the cases of equations of the second and third degrees there is an abundance of material. In the matter of notation Diophantus makes the greatest advance. He calls the coefficients of the unknown quantity πλῆθος. If there are several unknowns to be distinguished, he makes use of the ordinal numbers: ὁ πρῶτος ἀριθμός, ὁ δεύτερος, ὁ τρίτος. An equation† appears in his works in the abbreviated form:

---

*Matthiessen, *Grundzüge der antiken und modernen Algebra der litteralen Gleichungen*, 1878, p. 269. Hereafter referred to as Matthiessen.

† Matthiessen, p. 269.

κ˚βδ˚α ἴση ss^{ο˙˙ς}δ φ μ˚ιβ, i. e., $2x^3 + x^2 = 4x - 12$.

Diophantus classifies equations not according to the degree, but according to the number of essentially distinct terms. For this purpose he gives definite rules as to how equations can be brought to their simplest form, that is, the form in which both members of the equation have only positive terms. Practical problems which lead to equations of the first degree can be found in the works of Archimedes and Heron; the latter gives some of the so-called "fountain problems," which remind one of certain passages in the work of Ahmes. Equations of the second degree were mostly in the form of proportions, and this method of operation in the domain of a geometric algebra was well known to the Greeks. They undoubtedly understood how to represent by geometric figures equations of the form

$$\frac{a'}{a''}x = b, \quad \frac{a'}{a''}x + \frac{b'}{b''}y + \ldots = m,$$

where all quantities are linear. Every calculation of means in two equal ratios, i. e., in a proportion, was really nothing more than the solution of an equation. The Pythagorean school was acquainted with the arithmetic, the geometric and the harmonic means of two quantities; that is, they were able to solve geometrically the equations

$$x = \frac{a+b}{2}, \quad x^2 = ab, \quad x = \frac{2ab}{a+b}.$$

According to Nicomachus, Philolaus called the cube

with its six surfaces, its eight corners, and its twelve
edges, the geometric harmony, because it presented
equal measurements in all directions; from this fact,
it is said, the terms "harmonic mean" and "harmonic proportion" were derived, the relationship being :

$$\frac{12-8}{8-6} = \frac{12}{6}, \text{ whence } 8 = \frac{2 \cdot 6 \cdot 12}{6+12}, \text{ i. e., } x = \frac{2ab}{a+b}.$$

The number of distinct proportions was later increased to ten, although nothing essentially new was
gained thereby. Euclid gives thorough analyses of
proportions, that is, of the geometric solution of equations of the first degree and of incomplete quadratics ;
these, however, are not given as his own work, but as
the result of the labors of Eudoxus.

The solution of the equation of the second degree
by the geometric method of applying areas, largely
employed by the ancients, especially by Euclid, deserves particular attention.

In order to solve the equation

$$x^2 + ax = b^2$$

by Euclid's method, the problem must first be put in
the following form :

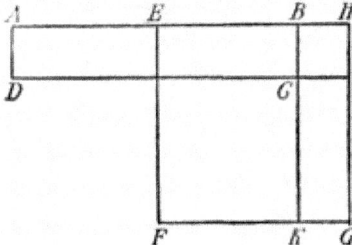

"To the segment $AB=a$ apply the rectangle DH of known area $=b^2$, in such a way that CH shall be a square." The figure shows that for $CK=\frac{a}{2}$, $FH=x^2+2x\cdot\frac{a}{2}+\left(\frac{a}{2}\right)^2=b^2+\left(\frac{a}{2}\right)^2$; but by the Pythagorean proposition, $b^2+\left(\frac{a}{2}\right)^2=c^2$, whence $EH=c=\frac{a}{2}+x$, from which we have $x=c-\frac{a}{2}$. The solution obtained by applying areas, in which case the square root is always regarded as positive, is accordingly nothing more than a constructive representation of the value

$$x=-\frac{a}{2}+\sqrt{b^2+\left(\frac{a}{2}\right)^2}=c-\frac{a}{2}.$$

In the same manner Euclid solves all equations of the form

$$x^2\pm ax\pm b^2=0,$$

and he remarks in passing that where $\sqrt{b^2-\left(\frac{a}{2}\right)^2}$, according to our notation, appears, the condition for a possible solution is $b>\frac{a}{2}$. Negative quantities are nowhere considered; but there is ground for inferring that in the case of two positive solutions the Greeks regarded both and that they also applied their method of solution to quadratic equations with numerical co-efficients.* By applying their knowledge of proportion, they were able to solve not only equations of the form $x^2\pm ax\pm b=0$, but also of the more general form

$$ax^2\pm ax\pm b^2=0,$$

for a as the ratio of two line-segments. Apollonius

---

*Zeuthen, *Die Lehre von den Kegelschnitten im Altertum.* Deutsch von v. Fischer-Benzon. 1886.

accomplished this with the aid of a conic, having the equation

$$y^2 = px \mp \tfrac{\ell}{a} x^2.$$

The Greeks were accordingly able to solve every general equation of the second degree having two essentially different coefficients, which might also contain numerical quantities, and to represent their positive roots geometrically.

The three principal forms of equations of the second degree first to be freed from geometric statement and completely solved, are

$$x^2 + px = q, \quad x^2 = px + q, \quad px = x^2 + q.$$

The solution consisted in applying an area, the problem being to apply to a given line a rectangle in such a manner that it would either contain a given area or be greater or less than this given area by a constant. For these three conditions there arose the technical expressions παραβολή, ὑπερβολή, ἔλλειψις, which after Archimedes came to refer to conics.*

In later times, with Heron and Diophantus, the solution of equations of the second degree was partly freed from the geometric representation, and passed into the form of an arithmetic computation proper (while disregarding the second sign in the square root).

The equation of the third degree, owing to its dependence on geometric problems, played an im-

* Tannery in *Bordeaux Mém.*, IV.

portant part among the Greeks. The problem of the
duplication (and also the multiplication) of the cube
attained especial celebrity. This problem demands
nothing more than the solution of the continued proportion $a:x=x:y=y:2a$, that is, of the equation
$x^3=2a^3$ (in general $x^3=\frac{m}{n}a^3$). This problem is very
old and was considered an especially important one
by the leading Greek mathematicians. Of this we
have evidence in a passage of Euripides in which he
makes King Minos say concerning the tomb of Glau-
cus which is to be rebuilt[*]: "The enclosure is too
small for a royal tomb: double it, but fail not in the
cubical form." The numerous solutions of the equation $x^3=2a^3$ obtained by Hippocrates, Plato, Me-
næchmus, Archytas and others, followed the geometric form, and in time the horizon was so considerably
extended in this direction that Archimedes in the
study of sections of a sphere solved equations of the
form

$$x^3-ax^2+b^2c=0$$

by the intersection of two lines of the second degree,
and in doing so also investigated the conditions to be
fulfilled in order that there should be no root or two
or three roots between 0 and a. Since the method
of reduction by means of which Archimedes obtains
the equation $x^3-ax^2+b^2c=0$ can be applied with
considerable ease to all forms of equations of the third
degree, the merit of having set forth these equations

* Cantor, I., p. 199.

in a comprehensive manner and of having solved one of their principal groups by geometric methods belongs without question to the Greeks.[*]

We find the first trace of indeterminate equations in the cattle problem (*Problema bovinum*) of Archimedes.

This problem, which was published in the year 1773 by Lessing, from a codex in the library at Wolfenbüttel, as the first of four unprinted fragments of Greek anthology, is given in twenty-two distichs. In all probability it originated directly with Archimedes who desired to show by means of this example how, proceeding from simple numerical quantities, one could easily arrive at very large numbers by the interweaving of conditions. The problem runs something as follows:[†]

The sun had a herd of bulls and cows of different colors. (1) Of Bulls the white (W) were in number $(\frac{1}{2} + \frac{1}{3})$ of the black (X) and the yellow (Y); the black (X) were $(\frac{1}{4} + \frac{1}{5})$ of the dappled (Z) and the yellow (Y); the dappled (Z) were $(\frac{1}{6} + \frac{1}{7})$ of the white (W) and the yellow (Y). (2) Of Cows which had the same colors (w, x, y, z). $w = (\frac{1}{3} + \frac{1}{4})(X + x)$, $x = (\frac{1}{4} + \frac{1}{5})(Z + z)$, $z = (\frac{1}{5} + \frac{1}{6})(Y + y)$, $y = (\frac{1}{6} + \frac{1}{7})(W + w)$. $W + X$ is to be a square; $Y + Z$ a triangular number.

The problem presents nine equations with ten unknowns:

$$W = (\tfrac{1}{2} + \tfrac{1}{3}) X + Y \qquad X = (\tfrac{1}{4} + \tfrac{1}{5}) Z + Y$$
$$Z = (\tfrac{1}{6} + \tfrac{1}{7}) W + Y \qquad w = (\tfrac{1}{3} + \tfrac{1}{4})(X + x)$$
$$x = (\tfrac{1}{4} + \tfrac{1}{5})(Z + z) \qquad z = (\tfrac{1}{5} + \tfrac{1}{6})(Y + y)$$
$$y = (\tfrac{1}{6} + \tfrac{1}{7})(W + w) \qquad W + X = n^2$$
$$Y + Z = \frac{m^2 + m}{2}.$$

[*] Zeuthen, *Die Lehre von den Kegelschnitten im Altertum.* Deutsch von v. Fischer-Benzon 1886.

[†] Krumbiegel und Amthor, *Das Problema bovinum des Archimedes.* Schlömilch's *Zeitschrift,* Bd. 25, Hl. A.; Gow, p. 99.

According to Amthor the solution is obtained by Pell's equation $t^2 - 2 \cdot 3 \cdot 7 \cdot 11 \cdot 29 \cdot 353\, u^2 = 1$, assuming the condition $u \equiv 0$ (mod. $2 \cdot 4657$), in which process there arises a continued fraction with a period of ninety-one convergents. If we omit the last two conditions, we get as the total number of cattle 5 916 837 175 686, a number which is nevertheless much smaller than that involved in the sand-reckoning of Archimedes.

But the name of Diophantus is most closely connected with systems of equations of this kind. He endeavors to satisfy his indeterminate equations not by means of whole numbers, but merely by means of rational numbers (always excluding negative quantities) of the form $\frac{p}{q}$ where p and q must be positive integers. It appears that Diophantus did not proceed in this field according to general methods, but rather by ingeniously following out special cases. At least those of his solutions of indeterminate equations of the first and second degrees with which we are acquainted permit of no other inference. Diophantus seems to have been not a little influenced by earlier works, such as those of Heron and Hypsicles. It may therefore be assumed that even before the Christian era there existed an indeterminate analysis upon which Diophantus could build.*

The Hindu algebra reminds us in many respects of Diophantus and Heron. As in the case of Diophantus, the negative roots of an equation are not admitted as solutions, but they are consciously set

* P. Tannery, in *Mémoires de Bordeaux*, 1880. This view of Tannery's is controverted by Heath, T. L., *Diophantos of Alexandria*, 1885, p. 135.

aside, which marks an advance upon Diophantus. The transformation of equations, the combination of terms containing the same powers of the unknown, is also performed as in the works of Diophantus. The following is the representation of an equation according to Bhaskara :*

$$\frac{va\ va\ 2\ \mid\ va\ 1\ \mid\ ru\ 30}{va\ va\ 0\ \mid\ va\ 0\ \mid\ ru\ 8}, \text{ i. e.,}$$

$2x^2 - x + 30 = 0x^2 + 0x + 8$, or $2x^2 - x + 30 = 8$.

Equations of the first degree appear not only with one, but also with several unknowns. The Hindu method of treating equations of the second degree shows material advance. In the first place, $ax^2 + bx = c$ is considered the only type† instead of the three Greek forms $ax^2 + bx = c$, $bx + c = ax^2$, $ax^2 + c = bx$. From this is easily derived $4a^2x^2 + 4abx = 4ac$, and then $(2ax + b)^2 = 4ac + b^2$, whence it follows that

$$x = \frac{-b + \sqrt{4ac + b^2}}{2a}.$$

Bhaskara goes still further. He considers both signs of the square root and also knows when it cannot be extracted. The two values of the root are, however, admitted by him as solutions only when both are positive,—evidently because his quadratic equations appear exclusively in connection with practical problems of geometric form. Bhaskara also solves equations of the third and fourth degrees in cases where these

* Matthiessen, p. 269.   † Cantor, I., p. 585.

equations can be reduced to equations of the second
degree by means of advantageous transformations and
the introduction of auxiliary quantities.

The indeterminate analysis of the Hindus is especially prominent. Here in contrast to Diophantus
only solutions in positive integers are admitted. Indeterminate equations of the first degree with two or
more unknowns had already been solved by Aryabhatta, and after him by Bhaskara, by a method in
which the Euclidean algorism for finding the greatest
common divisor is used; so that the method of solution corresponds at least in its fundamentals with the
method of continued fractions. Indeterminate equations of the second degree, for example those of the
form $xy = ax + by + c$, are solved by arbitrarily as-
signing a value to y and then obtaining x, or geometrically by the application of areas, or by a cyclic
method.* This cyclic method does not necessarily
lead to the desired end, but may nevertheless, by a
skilful selection of auxiliary quantities, give integral values. It consists in solving in the first place,
instead of the equation $ax^2 + b = cy^2$, the equation
$ax^2 + 1 = y^2$. This is done by the aid of the empiri-
cally assumed equation $aA^2 + B = C^2$, from which
other equations of the same form, $aA_n^2 + B_n = C_n^2$, can
be deduced by the solution of indeterminate equations
of the first degree. By means of skilful combinations

*Cantor, I., p. 591.

the equations $aA_n^2 + B_n = C_n^2$ furnish a solution of $ax^2 + 1 = y^2$.[*]

The algebra of the Chinese, at least in the earliest period, has this in common with the Greek, that equations of the second degree are solved geometrically. In later times there appears to have been developed a method of approximation for determining the roots of higher algebraic equations. For the solution of indeterminate equations of the first degree the Chinese developed an independent method. It bears the name of the "great expansion" and its discovery is ascribed to Sun tse, who lived in the third century A. D. This method can best be briefly characterised by the following example: Required a number x which when divided by 7, 11, 15 gives respectively the remainders 2, 5, 7. Let k_1, k_2, k_3. be found so that

$$\frac{11 \cdot 15 \cdot k_1}{7} = q_1 + \tfrac{1}{7}, \qquad \frac{15 \cdot 7 \cdot k_2}{11} = q_2 + \tfrac{1}{11},$$

$$\frac{7 \cdot 11 \cdot k_3}{15} = q_3 + \tfrac{1}{15};$$

we have, for example, $k_1 = 2$, $k_2 = 2$, $k_3 = 8$, and obtain the further results

$$11 \cdot 15 \cdot 2 = 330, \qquad 330 \cdot 2 = 660,$$
$$15 \cdot 7 \cdot 2 = 210, \qquad 210 \cdot 5 = 1050,$$
$$7 \cdot 11 \cdot 8 = 616, \qquad 616 \cdot 7 = 4312,$$
$$660 + 1050 + 4312 = 6022; \qquad \frac{6022}{7 \cdot 11 \cdot 15} = 5 + \frac{247}{7 \cdot 11 \cdot 15};$$

$x = 247$ is then a solution of the given equation.[†]

[*] Cantor, I., p. 593. [†] L. Matthiessen in *Schlömilch's Zeitschrift*, XXVI.

In the writing of their equations the Chinese make
as little use as the Hindus of a sign of equality. The
positive coefficients were written in red, the negative
in black. As a rule *täe* is placed beside the absolute
term of the equation and *yuen* beside the coefficient of
the first power ; the rest can be inferred from the example $14x^3 - 27x = 17,$[*] where r and b indicate the
color of the coefficient :

,14	or	,14	or	,14
,00		,00		,00
,27*yuen*		,27		,27*yuen*
,17*täe*		,17*täe*		,17.

The Arabs were pupils of both the Hindus and
the Greeks. They made use of the methods of their
Greek and Hindu predecessors and developed them,
especially in the direction of methods of calculation.
Here we find the origin of the word algebra in the
writings of Al Khowarazmi who, in the title of his
work, speaks of *"al-jabr wa'l muqabalah,"* i. e., the
science of redintegration and equation. This expression denotes two of the principal operations used by
the Arabs in the arrangement of equations. When
from the equation $x^3 + r = x^2 + px + r$ the new equation $x^3 = x^2 + px$ is formed, this is called *al muqabalah;*
the transformation which gives from the equation
$px - q = x^2$ the equation $px = x^2 + q$, a transformation which was considered of great importance by the

*Cantor, I., p. 643.

ancients, was called *al-jabr*, and this name was extended to the science which deals in general with equations.

The earlier Arabs wrote out their equations in words, as for example, Al Khowarazmi[*] (in the Latin translation):

Census et quinque radices equantur viginti quatuor

$$x^2 \quad + \quad 5x \quad\quad = \quad\quad 24;$$

and Omar Khayyam,

Cubus, latera et numerus aequales sunt quadratis

$$x^3 + bx + \quad c \quad\quad = \quad\quad ax^2.$$

In later times there arose among the Arabs quite an extended symbolism. This notation made the most marked progress among the Western Arabs. The unknown x was called *jidr*, its square *mal;* from the initials of these words they obtained the abbreviations $x = $ ش, $x^2 = $ و. Quantities which follow directly one after another are added, but a special sign is used to denote subtraction. "Equals" is denoted by the final letter of *adala* (equality), namely, by means of a final *lam*. In Al Kalsadi[†] $3x^2 = 12x + 63$ and $\frac{1}{2}x^2 + x = 7\frac{1}{2}$ are represented by

and the proportion $7 : 12 = 84 : x$ is given the form

$$\text{ج}. \cdot .84. \cdot .12. \cdot .7.$$

[*] Matthiessen, p. 269.　　　　[†] Cantor, I., p. 767.

Diophantus had already classified equations, not according to their degree, but according to the number of their terms. This principle of classification we find completely developed among the Arabs. According to this principle Al Khowarazmi* forms the following six groups for equations of the first and second degrees:

$x^2 = ax$ ("a square is equal to roots"),

$x^2 = a$ ("a square is equal to a constant"),

$ax = b$, $x^2 + ax = b$, $x^2 + a = bx$, $ax + b = x^2$,

("roots and a constant are equal to a square").

The Arabs knew how to solve equations of the first degree by four different methods, only one of which has particular interest, and that because in modern algebra it has been developed as a method of approximation for equations of higher degree. This method of solution, Hindu in its origin, is found in particular in Ibn al Banna and Al Kalsadi and is there called *the method of the scales.* It went over into the Latin translations as the *regula falsorum* and *regula falsi.* To illustrate, let the equation $ax + b = 0$ be given† and let z_1 and z_2 be any numerical quantities; then if we place $az_1 + b = y_1$, $az_2 + b = y_2$,

$$x = \frac{z_2 y_1 - z_1 y_2}{y_1 - y_2},$$

as can readily be seen. Ibn al Banna makes use of the following graphic plan for the calculation of the value of x:

* Matthiessen, p. 270.   † Matthiessen, p. 277.

The geometric representation, which with y as a negative quantity somewhat resembles a pair of scales, would be as follows, letting $OB_1 = z_1$, $OB_2 = z_2$, $B_1C_1 = y_1$, $B_2C_2 = y_2$, $OA = x$:

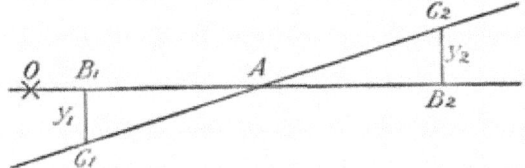

From this there results directly

$$\frac{x - z_1}{x - z_2} = \frac{y_1}{y_2};$$

that is, that the errors in the substitutions bear to each other the same ratio as the errors in the results, the method apparently being discovered through geometric considerations.

In the case of equations of the second degree Al Khowarazmi gives in the first place a purely mechanical solution (negative roots being recognised but not admitted), and then a proof by means of a geometric figure. He also undertakes an investigation of the number of solutions. In the case of

$$x^2 + c = bx, \text{ from which } x = \tfrac{b}{2} \pm \sqrt{(\tfrac{b}{2})^2 - c},$$

Al Khowarazmi obtains two solutions, one or none according as

$$(\tfrac{b}{2})^2 > c, \quad (\tfrac{b}{2})^2 = c, \quad (\tfrac{b}{2})^2 < c.$$

He gives the geometric proof for the correctness of the solution of an equation like $x^2 + 2x = 15$, where he takes $x = 3$, in two forms, either by means of a perfectly symmetric figure, or by the gnomon. In the first case, for $AB = x$, $BC = \tfrac{1}{2}$, $BD = 1$, we have

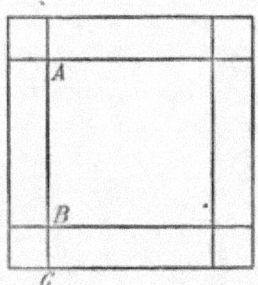 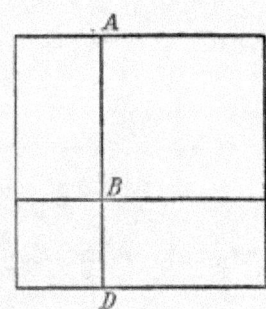

$x^2 + 4 \cdot \tfrac{1}{2} \cdot x + 4 \cdot (\tfrac{1}{2})^2 = 15 + 1$, $(x+1)^2 = 16$; in the second we have $x^2 + 2 \cdot 1 \cdot x + 1^2 = 15 + 1$. In the treatment of equations of the form $ax^{2n} \pm bx^n \pm c = 0$, the theory of quadratic equations receives still further development at the hands of Al Kalsadi.

Equations of higher degree than the second, in the form in which they presented themselves to the Arabs in the geometric or stereometric problems of the Greek type, were not solved by them arithmetically, but only by geometric methods with the aid of conics. Here Omar Khayyam * proceeded most systematically. He solved the following equations of the third degree geometrically:

* Matthiessen, pp. 367, 894, 945.

$$r = x^3, \quad x^3 \pm px^2 = qx, \quad x^3 + r = qx, \quad x^3 \pm px^2 \pm qx = r,$$
$$qx = x^3, \quad x^3 + qx = px^2, \quad x^3 \pm px^2 = r, \quad x^3 \pm px^2 + r = qx,$$
$$px^2 = x^3, \quad x^3 \pm qx = r, \quad x^3 + r = px^2, \quad x^3 \pm px^2 = qx + r.$$

The following is the method of expression which he employs in these cases:

"A cube and square are equal to roots;"
"a cube is equal to roots, squares and one number,"
when the equations

$$x^3 + px^2 = qx, \quad x^3 = px^2 + qx + r$$

are to be expressed. Omar calls all binomial forms simple equations; trinomial and quadrinomial forms he calls composite equations. He was unable to solve the latter, even by geometric methods, in case they reached the fourth degree.

The indeterminate analysis of the Arabs must be traced back to Diophantus. In the solution of indeterminate equations of the first and second degree Al Karkhi gives integral and fractional numbers, like Diophantus, and excludes only irrational quantities. The Arabs were familiar with a number of propositions in regard to Pythagorean triangles without having investigated this field in a thoroughly systematic manner.

C. THE SECOND PERIOD.

TO THE MIDDLE OF THE SEVENTEENTH CENTURY.

As long as the cultivation of the sciences among the Western peoples was almost entirely confined to the monasteries, during a period lasting from the eighth century to the twelfth, no evidence appeared of any progress in the general theory of numbers. As in the learned Roman world after the end of the fifth century, so now men recognised seven liberal arts,—the *trivium*, embracing grammar, rhetoric and dialectics, and the *quadrivium*, embracing arithmetic, geometry, music and astronomy.* But through Arab influence, operating in part directly and in part through writings, there followed in Italy and later also in France and Germany a golden age of mathematical activity whose influence is prominent in all the literature of that time. Thus Dante, in the fourth canto of the *Divina Commedia* mentions among the personages

> ". . . who slow their eyes around
> Majestically moved, and in their port
> Bore eminent authority,"

a Euclid, a Ptolemy, a Hippocrates and an Avicenna.

There also arose, as a further development of certain famous cloister, cathedral and chapter schools, and in rare instances, independent of them, the first universities, at Paris, Oxford, Bologna, and Cambridge, which in the course of the twelfth century associated the separate faculties, and from the beginning of the thirteenth century became famous as *Studia generalia*.† Before long universities were also es-

---

* Müller, *Historisch-etymologische Studien über mathematische Terminologie*, 1887.

† Suter, *Die Mathematik auf den Universitäten des Mittelalters*, 1887.

tablished in Germany (Prague, 1348; Vienna, 1365; Heidelberg, 1386; Cologne, 1388; Erfurt, 1392; Leipzig, 1409; Rostock, 1419; Greifswalde, 1456; Basel, 1459; Ingolstadt, 1472; Tübingen and Mainz, 1477), in which for a long while mathematical instruction constituted merely an appendage to philosophical research. We must look upon Johann von Gmunden as the first professor in a German university to devote himself exclusively to the department of mathematics. From the year 1420 he lectured in Vienna upon mathematical branches only, and no longer upon all departments of philosophy, a practice which was then universal.

1. *General Arithmetic.*

Even Fibonacci made use of words to express mathematical rules, or represented them by means of line-segments. On the other hand, we find that Luca Pacioli, who was far inferior to his predecessor in arithmetic inventiveness, used the abbreviations *.p.*, *.m.*, *R.* for *plus*, *minus*, and *radix* (root). As early as 1484, ten years before Pacioli, Nicolas Chuquet had written a work, in all probability based upon the researches of Oresme, in which there appear not only the signs $\bar{p}$ and $\bar{m}$ (for *plus* and *minus*), but also expressions like

$$R^4.10, \quad R^2.17 \text{ for } \sqrt[4]{10}, \sqrt{17}.$$

He also used the Cartesian exponent-notation, and

the expressions *equipolence, equipolent,* for equivalence and equivalent.[*]

Distinctively symbolic arithmetic was developed upon German soil. In German general arithmetic and algebra, in the *Deutsche Coss,* the symbols $+$ and $-$ for plus and minus are characteristic.[†] They were in common use while the Italian school was still writing $\bar{p}$ and $\bar{m}$. The earliest known appearance of these signs is in a manuscript (*Regula Cose vel Algebre*) of the Vienna library, dating from the middle of the fifteenth century. In the beginning of the seventeenth century Reymers and Faulhaber used the sign $\div$,[‡] and Peter Roth the sign $\ddagger$ as minus signs.

Among the Italians of the thirteenth and fourteenth centuries, in imitation of the Arabs, the course of an arithmetic operation was expressed entirely in words. Nevertheless, abbreviations were gradually introduced and Luca Pacioli was acquainted with such abbreviations to express the first twenty-nine powers of the unknown quantity. In his treatise the absolute term and x, x^2, x^3, x^4, x^5, x^6, . . . are always respectively represented by *numero* or n^o, *cosa* or *co, censo* or *ce, cubo* or *cu, censo de censo* or *ce.ce, primo relato* or *p.^{o}r^o, censo de cuba* or *ce.cu* . . .

The Germans made use of symbols of their own

[*] A. Marre in *Boncompagni's Bulletino,* XIII. *Jahrbuch über die Fortschritte der Math.,* 1881, p. 8.

[†] Treutlein, "Die deutsche Coss," *Schlömilch's Zeitschrift,* Bd. 24, Hl. A. Hereafter referred to as Treutlein.

[‡] The sign $\div$ is first used as a sign of division in Rahn's *Teutsche Algebra* Zürich, 1659.

invention. Rudolff and Riese represented the absolute term and the powers of the unknown quantity in
the following manner: *Dragma*, abbreviated in writing, ϕ; *radix* (or *coss*, i. e., *root* of the equation) is
expressed by a sign resembling an *r* with a little flourish; *zensus* by $\mathfrak{z}$; *cubus* by *c* with a long flourish on
top in the shape of an *l* (in the following pages this
will be represented merely by *c*); *zensus de zensu* (zens-
dezens) by $\mathfrak{zz}$, *sursolidum* by β or $\mathfrak{f}$; *zensikubus* by $\mathfrak{z}c$;
bissursolidum by $\mathrm{bi}\mathfrak{f}$ or $\mathrm{B}\mathfrak{f}$; *zensus zensui de zensu* (zens-
zensdezens) by $\mathfrak{zzz}$; *cubus de cubo* by *cc*.

There are two opinions concerning the origin of
the *x* of mathematicians. According to the one, it was
originally an *r* (*radix*) written with a flourish which
gradually came to resemble an *x*, while the original
meaning was forgotten, so that half a century after
Stifel it was read by all mathematicians as *x*.[*] The
other explanation depends upon the fact that it is customary in Spain to represent an Arabic *s* by a Latin
x where whole words and sentences are in question;
for instance the quantity 12*x*, in Arabic شِ
is repre-
sented by 12 *xai*, more correctly by 12 *sai*. According to this view, the *x* of the mathematicians would
be an abbreviation of the Arabic *sai* $=$ *xai*, an expression for the unknown quantity.

By the older cossists[†] these abbreviations are introduced without any explanation; Stifel, however,

[*] Treutlein. G. Wertheim in *Schlömilch's Zeitschrift*, Bd. 44, Hf. A.
[†] Treutlein.

considers it necessary to give his readers suitable explanations. The word "root," used for the first power
of an unknown quantity he explains by means of
the geometric progression, "because all successive
members of the series develop from the first as from
a root"; he puts for x^0, x^1, x^2, x^3, x^4, ... the signs 1,
$1x$, 1_3, $1c$, 1_{33}, ... and calls these "cossic numbers,"
which can be continued to infinity, while to each is
assigned a definite order-number, that is, an exponent. In the German edition of Rudolff's *Coss*, Stifel
at first writes the "cossic series" to the seventeenth
power in the manner already indicated, but also later
as follows:

$$\overset{0}{1} \ . \ \overset{1}{1\mathfrak{A}.} \quad \overset{2}{1\mathfrak{A}\mathfrak{A}.} \quad \overset{3}{1\mathfrak{A}\mathfrak{A}\mathfrak{A}.} \quad \overset{4}{1\mathfrak{A}\mathfrak{A}\mathfrak{A}\mathfrak{A}.} \quad \text{etc.}$$

He also makes use of the letters $\mathfrak{B}$ and $\mathfrak{C}$ in writing
this expression. The nearest approach to our present
notation is to be found in Bürgi and Reymers, where
with the aid of "exponents" or "characteristics" the
polynomial $8x^6 + 12x^5 - 9x^4 + 10x^3 + 3x^2 + 7x - 4$ is
represented in the following manner:

$$\overset{\text{vi}}{8} + \overset{\text{v}}{12} - \overset{\text{iv}}{9} + \overset{\text{iii}}{10} + \overset{\text{ii}}{3} + \overset{\text{i}}{7} - \overset{\text{o}}{4}$$

In Scheubel we find for x, x^2, x^3, x^4, x^5 ..., *pri.*,
sec., *ter.*, *quar.*, *quin.*, and in Ramus *l*, *q*, *c*, *bq*, *s* as
abbreviations for *latus, quadratus, cubus, biquadratus,
solidus.*

The product $(7x^2 - 3x + 2)(5x - 3) = 35x^3 - 36x^2 + 19x - 6$ is represented in its development by Grammateus, Stifel, and Ramus in the following manner:

GRAMMATEUS:

$$7x. -3\,pri. + 2N$$
$$\text{by} \quad 5\,pri. - 3N$$
$$35\,ter. -15x. + 10\,pri.$$
$$-21x. + 9\,pri. - 6N$$
$$35\,ter. -36x. + 19\,pri. - 6N$$

STIFEL:

$$7_3 -3x +2$$
$$5x -3$$
$$35\,c -15_3 + 10x$$
$$-21_3 + 9x - 6$$
$$35\,c -36_3 + 19x - 6$$

RAMUS:

$$7q- 3l + 2$$
$$5l - 3$$
$$35c -15q + 10l$$
$$-21q + 9l - 6$$
$$35c - 36q + 19l - 6.$$

As early as the fifteenth century the German Coss made use of a special symbol to indicate the extraction of the root. At first .4 was used for $\sqrt{4}$; this period placed before the number was soon extended by means of a stroke appended to it. Riese and Rudolff write merely $\sqrt{}4$ for $\sqrt{4}$. Stifel takes the first step towards a more general comprehension of radical quantities in his *Arithmetica integra,* where the second, third, fourth, fifth, roots of six are represented by $\sqrt{}3^6$, $\sqrt{}c^6$, $\sqrt{}33^6$, $\sqrt{}\frac{c}{3}^6$, while elsewhere the symbols

are used as radical signs. These symbols, of which the first two occur in Rudolff and the other three in a work of Stifel, indicate respectively the third, fourth, second, third, and fourth roots of the numbers which they precede.

Rudolff gives a few rules for operations with radical quantities, but without demonstrations. Like Fibonacci he calls an irrational number a *numerus surdus*. Such expressions as the following are introduced:

$$\sqrt{a} \pm \sqrt{b} = \sqrt{a + b \pm \sqrt{4ab}},$$
$$\sqrt{a^2 c} + \sqrt{b^2 c} = \sqrt{(a + b)^2 \cdot c},$$
$$\frac{x}{\sqrt{a} \pm \sqrt{b}} = \frac{x\left(\sqrt{a} \mp \sqrt{b}\right)}{a - b}.$$

Stifel enters upon the subject of irrational numbers with especial zeal and even refers to the speculations of Euclid, but preserves in all his developments a well-grounded independence. Stifel distinguishes two classes of irrational numbers: principal and subordinate (*Haupt- und Nebenarten*). In the first class are included (1) simple irrational numbers of the form $\sqrt[i]{a}$, (2) binomial irrationals with the positive sign, as

$$\sqrt{33^{10}} + \sqrt{33^6}, \quad 4 + \sqrt{3^6}, \quad \sqrt{3^{12}} + \sqrt{c^{12}},$$

(3) square roots of such binomial irrationals as

$$\sqrt{3 \cdot \sqrt{3^6}} + \sqrt{3^8} = \sqrt{\sqrt{6} + \sqrt{8}};$$
$$\sqrt{3 \cdot 5} + \sqrt{3^5} = \sqrt{5 + \sqrt{5}},$$

(4) binomial irrationals with the negative sign, as $\sqrt{33^{10}} - \sqrt{33^6}$, and (5) square roots of such binomial irrationals, as

$$\sqrt{3 \cdot \sqrt{3^6}} - \sqrt{3^8} = \sqrt{\sqrt{6} - \sqrt{8}}.$$

The subordinate class of irrational quantities, according to Stifel, includes expressions like

$$\sqrt{3^2} + \sqrt{3^3} + \sqrt{3^5}, \quad \sqrt{3^2} + \sqrt{33^4} + \sqrt{c^3},$$
$$\sqrt{33} \cdot \sqrt{3^6} + 2. - \cdot \sqrt{3c} \cdot \sqrt{c^8} + \sqrt{33^{12}}$$
$$= \sqrt[l]{\sqrt{6} + 2} - \sqrt[l]{\sqrt[3]{8} + \sqrt[l]{12}}$$

Fibonacci evidently obtained his knowledge of negative quantities from the Arabs, and like them he does not admit negative quantities as the roots of an equation. Pacioli enunciates the rule, minus times minus gives always plus, but he makes use of it only for the expansion of expressions of the form $(p-q)$ $(r-s)$. Cardan proceeds in the same way; he recognises negative roots of an equation, but he calls them *aestimationes falsae* or *fictae*,* and attaches to them no independent significance. Stifel calls negative quantities *numeri absurdi*. Harriot is the first to consider negative quantities in themselves, allowing them to form one side of an equation. Calculations involving negative quantities consequently do not begin until the seventeenth century. It is the same with irrational numbers; Stifel is the first to include them among numbers proper.

Imaginary quantities are scarcely mentioned. Cardan incidentally proves that

$$\left(5 + \sqrt{-15}\right) \cdot \left(5 - \sqrt{-15}\right) = 40.$$

Bombelli goes considerably farther. Although not entering into the nature of imaginary quantities, of which he calls $+\sqrt{-1}$ *piu di meno*, and $-\sqrt{-1}$ *meno di meno*, he gives rules for the treatment of ex-

* *Ars magna*, 1545. Cap. I., 6.

pressions of the form $a + b\sqrt{-1}$, as they occur in the solution of the cubic equation.

The Italian school early made considerable advancement in calculations involving powers. Nicole Oresme[*] had long since instituted calculations with fractional exponents. In his notation

$$\tfrac{1}{2} \cdot 1'\tfrac{1}{3} = (1\tfrac{1}{3})^{\frac{1}{2}}, \quad \tfrac{1}{4} \cdot 2'\tfrac{1}{2} = (2\tfrac{1}{2})^{\frac{1}{4}}$$

it appears that he was familiar with the formulæ

$$a^{\frac{m}{n}} = (a^m)^{\frac{1}{n}}, \quad a^{\frac{1}{m}} \cdot \beta^{\frac{1}{n}} = (a^n \cdot \beta^m)^{\frac{1}{mn}}, \quad a^{\frac{1}{m}} : \beta^{\frac{1}{n}} = \left(\frac{a^n}{\beta^m}\right)^{\frac{1}{mn}}.$$

In the transformation of roots Cardan made the first important advance by writing

$$\sqrt[v]{a + \sqrt{b}} = p + \sqrt{q}, \quad \sqrt[v]{a - \sqrt{b}} = p - \sqrt{q}$$

and therefore $\sqrt[v]{a^2 - b} = p^2 - q = c$, $a^2 - b = c^3$. Bombelli[†] enlarged upon this observation and wrote

$$\sqrt[v]{a + \sqrt{-b}} = p + \sqrt{-q}, \quad \sqrt[v]{a - \sqrt{-b}} = p - \sqrt{-q},$$

from which follows $\sqrt[v]{a^2 + b} = p^2 + q$. With reference to the equation $x^3 = 15x + 4$ he discovered that

$$x = \sqrt[v]{2 + \sqrt{-121}} + \sqrt[v]{2 - \sqrt{-121}}$$
$$= 2 + \sqrt{-1} + 2 - \sqrt{-1} = 4.$$

For in this case

$$p^2 + q = 5, \quad (p + \sqrt{-q})^3 = 2 + \sqrt{-121},$$
$$(p - \sqrt{-q})^3 = 2 - \sqrt{-121},$$

become through addition $p^3 - 3pq = 2$, and with $q = 5 - p^2$, $4p^3 - 15p = 2$, and consequently (by trial) $p = 2$ and $q = 1$.

* Hankel, p. 350.   † Cantor, II., p. 572.

The extraction of square and cube roots according to the Arab, or rather the Indian, method, was set forth by Grammateus. In the process of extracting the square root, for the purpose of dividing the number into periods, points are placed over the first, third, fifth, etc., figures, counting from right to left. Stifel[*] developed the extracting of roots to a considerable extent; it is undoubtedly for this purpose that he worked out a table of binomial coefficients as far as $(a + b)^{17}$, in which, for example, the line for $(a + b)^4$ reads:

$$1_{33} \quad . \quad 4 \quad 6 \quad 4 \quad 1 \quad .$$

The theory of series in this period made no advance upon the knowledge of the Arabs. Peurbach found the sum of the arithmetic and the geometric progressions. Stifel examined the series of natural numbers, of even and of odd numbers and deduced from them certain power series. In regard to these series he was familiar, through Cardan, with the theorem that $1 + 2 + 2^2 + 2^3 + .. + 2^{n-1} = 2^n - 1$. With Stifel geometric progressions appear in an application which is not found in Euclid's treatment of means.[†] As is well known, n geometric means are inserted between the two quantities a and b by means of the equations

$$\frac{a}{x_1} = \frac{x_1}{x_2} = \frac{x_2}{x_3} = \ldots = \frac{x_{n-1}}{x_n} = \frac{x_n}{b} = q$$

[*] Cantor, II., pp. 397, 409.   [†] Treutlein.

where $q = {}^{n+1}\!\sqrt{\tfrac{a}{b}}$.   Stifel inserts five geometric means between the numbers 6 and 18 in the following manner:

1	3	9	27	81	243	729
$\sqrt[3]{3}\cdot1$	$\sqrt[3]{3}\cdot3$	$\sqrt[3]{3}\cdot9$	$\sqrt[3]{3}\cdot27$	$\sqrt[3]{3}\cdot81$	$\sqrt[3]{3}\cdot243$	$\sqrt[3]{3}\cdot729$
6	$\sqrt[3]{3}\cdot139968$	$\sqrt[3]{648}$	$\sqrt[3]{108}$	$\sqrt[3]{1944}$	$\sqrt[3]{3}\cdot11337408$	18

in which the last line is obtained from the preceding by multiplying by 6.   Stifel makes use of this solution for the purpose of duplicating the cube.   He selects 6 for the edge of the given cube; three geometric means are to be inserted between 6 and 12, and as $q = \sqrt[3]{\tfrac{1}{2}}$, the edge of the required cube will be $x = 6\sqrt[3]{2} = \sqrt[3]{432}$.   This length is constructed geometrically by Stifel in the following manner:

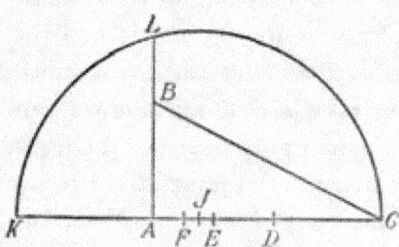

In the right angled triangle ACB, with the hypotenuse BC, let $AB = 6$, $AC = 12$; make $AD = DC$, $AE = ED$, $AF = FE$, $FJ = JE$, $JK = JC = JL$.   Then AK is the first, AL the second geometric mean between 6 and 12.   This construction, which Stifel regards as entirely correct, is only an approximation, since $AK = 7.5$ instead of $6\sqrt[3]{2} = 7.56$, $AL = 3\sqrt{10} = 9.487$ instead of $6\sqrt[3]{4} = 9.524$.

Simple facts involving the theory of numbers were also known to Stifel, such as theorems relating to perfect and diametral numbers and to magic squares.

A diametral number is the product of two numbers the sum of whose squares is a rational square, the square of the diameter, e. g., $65^2 = 25^2 + 60^2 = 39^2 + 52^2$, and hence $25.60 = 1500$ and $39.52 = 2028$ are diametral numbers of equal diameter.

11	24	7	20	3
4	12	25	8	16
17	5	13	21	9
10	18	1	14	22
23	6	19	2	15

Magic squares are figures resembling a chess board, in which the terms of an arithmetic progression are so arranged that their sum, whether taken diagonally or by rows or columns, is always the same. A magic square containing an odd number of cells, which is easier to construct than one containing an even number, can be obtained in the following manner: Place 1 in the cell beneath the central one, and the other numbers, in their natural order, in the empty cells in a diagonal direction. Upon coming to a cell already occupied, pass vertically downwards over two cells.* Possibly magic squares were known to the Hindus; but of this there is no certain evidence.† Manuel Moschopulus‡ (probably in the fourteenth century) touched upon the subject of magic squares. He gave definite rules for the construction of these figures, which long after found a wider diffusion through Lahire and Mollweide. During the Middle Ages magic squares formed a part of the wide-spread number-mysticism. Stifel was the first to investigate them in a scientific way, although Adam Riese had already introduced the subject into Germany, but neither he nor Riese was able to give a simple rule for their construction. We may nevertheless assume that towards the end of the sixteenth century such rules were known to a few German mathematicians,§ as for instance, to the *Rechenmeister* of Nuremberg, Peter Roth. In the year 1612 Bachet published in his *Problèmes plaisants* ‖ a general rule for squares containing an odd number of cells, but acknowledged that he had not succeeded in finding a solution for squares containing an even number. Frénicle was the first to make a real advance beyond Bachet. He gave rules (1693) for both classes of squares, and even discovered squares that maintain their characteristics after striking off th

* Unger, p. 109.
† Montucla, *Histoire des Mathématiques*, 1799-1802.
‡ Cantor, I., p. 480.
§ Giesing, *Leben und Schriften Leonardo's da Pisa*, 1886.
‖ This work is now accessible in a new edition published in 1884, Paris, Gauthier-Villars.

outer rows and columns. In 1816 Mollweide collected the scattered rules into a book, *De quadratis magicis*, which is distinguished by its simplicity and scientific form. More modern works are due to Hugel (Ansbach, 1859), to Pessl (Amberg, 1872), who also considers a magic cylinder, and to Thompson (*Quarterly Journal of Mathematics*, Vol. X.), by whose rules the magic square with the side pn is deduced from the square with the side n.[*]

2. *Algebra.*

Towards the end of the Middle Ages the *Ars major*, *Arte maggiore*, *Algebra* or the *Coss* is opposed to the ordinary arithmetic (*Ars minor*). The Italians called the theory of equations either simply *Algebra*, like the Arabs, or *Ars magna*, *Ars rei et census* (very common after the time of Leonardo and fully settled in Regiomontanus), *La regola della cosa* (*cosa = res*, *thing*), *Ars cossica* or *Regula cosae*. The German algebraists of the fifteenth and sixteenth centuries called it *Coss*, *Regula Coss*, *Algebra*, or, like the Greeks, *Logistic*. Vieta used the term *Arithmetica speciosa*, and Reymers *Arithmetica analytica*, giving the section treating of equations the special title *von der Aequation*. The method of representing equations gradually took on the modern form. Equality was generally, even by the cossists, expressed by words; it was not until the middle of the seventeenth century that a special sym-

[*] Günther, "Ueber magische Quadrate," *Grunert's Arch.*, Bd. 57.

bol came into common use.   The following are examples of the different methods of representing equations :*

Cardan :

   Cubus p 6 rebus aequalis 20, $x^3 + 6x = 20$;

Vieta :

   $1C - 8Q + 16N$ aequ. 40, $x^3 - 8x^2 + 16x = 40$;

Regiomontanus :

   16 census et 2000 aequ. 680 rebus, $16x^2 + 2000 = 680x$;

Reymers :

XXVIII XII		X	VI	III	I	O
$1gr$ 65532	$+18$	$\div 30$	$\div 18$	$+12$	$\div 8$;

$$x^{28} = 65532x^{12} + 18x^{10} - 30x^6 - 18x^3 + 12x - 8;$$

Descartes :

$z^2 \infty\ az - bb$ $z^2 = az - b^2$;

$y^4 - 8y^3 - 1yy + 8y^* \infty\ 0,$ $y^4 - 8y^3 - y^2 + 8y = 0$;

$x^6\ *\ *\ *\ * - bx \infty 0,$ $x^6 - bx$ $= 0$;

$x^5\ *\ *\ *\ * - b\ \ \infty 0,$ $x^5 - b$ $= 0$;

Hudde :

   $x^3 \infty\ qx.r,$ $x^3 = qx + r.$

In Euler's time the last transformation in the development of the modern form had already been accomplished.

Equations of the first degree offer no occasion for remark.   We may nevertheless call attention to the peculiar form of the proportion which is found in Grammateus and Apian.† The former writes : "Wie

*Matthiessen, *Grundzüge der antiken und modernen Algebra*, 2 ed., 1896, p. 270, etc.

†Gerhardt, *Geschichte der Mathematik in Deutschland*, 1877.

sich hadt *a* zum *b*, also hat sich *c* zum *d*," and the latter places

$$4-12-9-0 \text{ for } \frac{4}{12} = \frac{9}{x}.$$

Leonardo of Pisa solved equations of the second degree in identically the same way as the Arabs.* Cardan recognized two roots of a quadratic equation, even when one of them was negative; but he did not regard such a root as forming an actual solution. Rudolff recognized only positive roots, and Stifel stated explicitly that, with the exception of the case of quadratic equations with two positive roots, no equation can have more than one root. In general, the solution was affected in the manner laid down by Grammateus† in the example $12x + 24 = 2\frac{10}{49}x^2$: "Proceed thus: divide $24N$ by $2\frac{10}{49}$ sec., which gives $10\frac{8}{9}a$ ($10\frac{8}{9} = a$). Also divide 12 pri. by $2\frac{10}{49}$ sec., which gives the result $5\frac{4}{9}b$ ($5\frac{4}{9} = b$). Square the half of b, which gives $\frac{2401}{324}$, to which add $a = 10\frac{8}{9}$, giving $\frac{5929}{324}$, of which the square root is $\frac{77}{18}$. Add this to $\frac{1}{2}$ of b, or $\frac{49}{18}$, and 7 is the number represented by 1 pri. Proof: $12 \times 7N = 84N$; add $24N$, $= 108N$. $2\frac{10}{49}$ sec. multiplied by 49 must also give $108N$."

This "German Coss" was certainly cultivated by Hans Bernecker in Leipzig and by Hans Conrad in Eisleben‡ (about 1525), yet no memoranda by either of these mathematicians have been found. The University of Vienna encouraged Grammateus to publish,

---

*Cantor, II., p. 31. † Gerhardt. ‡ Cantor, II., p. 387.

in the year 1523, the first German treatise on Algebra
under the title, "*Eyn new kunstlich behend vnd gewiss
Rechenbüchlin | vff alle Kauffmannschafft.* Nach Ge-
meynen Regeln de tre. Welschen practic. Regeln
falsi. Etlichen Regeln Cosse . . Buchhalten . . Visier
Ruthen zu machen." Adam Riese, who had published his Arithmetic in 1518, completed in 1524 the
manuscript of the Coss; but it remained in manuscript and was not found until 1855 in Marienberg.
The Coss published by Christoff Rudolff in 1525 in
Strassburg met with universal favor. This work,
which is provided with many examples, all completely
solved, is described in the following words:

"Behend vnd Hübsch Rechnung durch die kunstreichen re-
geln Algebre | so gemeinicklich die Coss genennt werden. Darinnen
alles so treulich an Tag geben | das auch allein ausz vleissigem
Lesen on allen mündtliche vnterricht mag begriffen werden. Hind-
angesetzt die meinung aller dere | so bisher vil vngegründten regeln
angehangen. Einem jeden liebhaber diser kunst lustig vnd ergetz-
lich Zusamen bracht durch Christoffen Rudolff von Jawer."*

The principal work of the German Coss is Michael
Stifel's *Arithmetica integra*, published in Nuremberg in
1544. In this book, besides the more common operations of arithmetic, not only are irrational quantities
treated at length, but there are also to be found appli-

---

*A translation would read somewhat as follows: "Rapid and neat com-
putation by means of the ingenious rules of algebra, commonly designated
the Coss. Wherein are faithfully elucidated all things in such wise that they
may be comprehended from diligent reading alone, without any oral instruction whatsoever. In disregard of the opinions of all those who hitherto have
adhered to numerous unfounded rules. Happily and divertingly collected
'or lovers of this art, by Christoff Rudolff, of Jauer."

cations of algebra to geometry. Stifel also published
in 1553 *Die Coss Christoffs Rudolffs mit schönen Ex-
empeln der Coss Gebessert vnd sehr gemehrt*, with copi-
ous appendices of his own, giving compendia of the
Coss. With pardonable self-appreciation Stifel asserts, "It is my purpose in such matters (as far as I
am able) from complexity to produce simplicity.
Therefore from many rules of the Coss I have formed
a single rule and from the many methods for roots
have also established one uniform method for the in-
numerable cases."

Stifel's writings were laid under great contribution by later writers on mathematics in widely distant
lands, usually with no mention of his name. This
was done in the second half of the sixteenth century
by the Germans Christoph Clavius and Scheubel, by
the Frenchmen Ramus, Peletier, and Salignac, by
the Dutchman Menher, and by the Spaniard Nuñez.
It can, therefore, be said that by the end of the sixteenth century or the beginning of the seventeenth
the spirit of the German Coss dominated the Algebra
of all the European lands, with the single exception
of Italy.

The history of the purely arithmetical solution of
equations of the third and fourth degrees which was
successfully worked out upon Italian soil demands
marked attention. Fibonacci (Leonardo of Pisa)*
made the first advance in this direction in connection

* Cantor, II., p. 43.

with the equation $x^3 + 2x^2 + 10x = 20$. Although he succeeded in solving this only approximately, it furnished him with the opportunity of proving that the value of x cannot be represented by square roots alone, even when the latter are chosen in compound form, like

$$\sqrt{\sqrt{m} \pm \sqrt{n}}.$$

The first complete solution of the equation $x^3 + mx = n$ is due to Scipione del Ferro, but it is lost.* The second discoverer is not Cardan, but Tartaglia. On the twelfth of February, 1535, he gave the formula for the solution of the equation $x^3 + mx = n$, which has since become so famous under the name of his rival. By 1541 Tartaglia was able to solve any equation whatsoever of the third degree. In 1539 Cardan enticed his opponent Tartaglia to his house in Milan and importuned him until the latter finally confided his method under the pledge of secrecy. Cardan broke his word, publishing Tartaglia's solution in 1545 in his *Ars magna*, although not without some mention of the name of the discoverer. Cardan also had the satisfaction of giving to his contemporaries, in his *Ars magna*, the solution of the biquadratic equation which his pupil Ferrari had succeeded in obtaining. Bombelli is to be credited with representing the roots of the equation of the third degree in the simplest form, in the so-called irreducible case, by means of a transformation of the irrational quantities. Of the German

* Hankel, p. 360.

mathematicians, Rudolff also solved a few equations of the third degree, but without explaining the method which he followed. Stifel by this time was able to give a brief account of the "cubicoss," that is, the theory of equations of the third degree as given in Cardan's work. The first complete exposition of the Tartaglian solution of equations of the third degree comes from the pen of Faulhaber (1604).

The older cossists* had arranged equations of the first, second, third, and fourth degrees (in so far as they allow of a solution by means of square roots alone) in a table containing twenty-four different forms. The peculiar form of these rules, that is, of the equations with their solutions, can be seen in the following examples taken from Riese:

"The first rule is when the root [of the equation] is equal to a number, or dragma so called. Divide by the number of roots; the result of this division must answer the question." (I. e., if $ax = b$, then

$$x = \frac{b}{a}.$$

"The sixteenth rule is when squares equal cubes and fourth powers. Divide through by the number of fourth powers [the coefficient of x^4], then take half the number of cubes and multiply this by itself, add this product to the number of squares, extract the square root, and from the result take half the number of cubes. Then you have the answer."

* Treutlein.

Taking this step by step we have,

$$ax^4 + bx^3 = cx^2, \quad x^4 + \frac{b}{a}x^3 = \frac{c}{a}x^2, \text{ or}$$

$$x^4 + ax^3 = \beta x^2, \quad x = \sqrt{\left(\frac{a}{2}\right)^2 + \beta} - \frac{a}{2}.$$

The twenty-four forms of the older cossists are reduced by Riese to "acht equationes" (eight equations, as his combination of German and Latin means),
but as to the fact that the square root is two-valued
he is not at all clear. Stifel was the first to let a single
equation stand for these eight, and he expressly asserts that a quadratic can have only two roots; this
he asserts, however, only for the equation $x^2 = ax - b$.
In order to reduce the equations above mentioned to
one of Riese's eight forms, Rudolff availed himself of
"four precautions (Cautelen)," from which it is clearly
seen what labor it cost to develop the coss step by
step. For example, here is his

"First precaution. When in equating two numbers, in the one is found a quantity, and in the other
is found one of the same name, then (considering the
signs + and —) must one of these quantities be added
to or subtracted from the other, one at a time, care
being had to make up for the defect in the equated
numbers by subtracting the + and adding the —."
(I. e., from $5x^2 - 3x + 4 = 2x^2 + 5x$, we derive $3x^2 + 4 = 8x$.)

The first examples of this period, of equations
with more than one unknown quantity, are met with

in Rudolff,* who treats them only incidentally. Here also Stifel went decidedly beyond his predecessors. Besides the first unknown, $1x$, he introduced $1A$, $1B$, $1C$, . . . as *secundae radices* or additional unknowns and indicated the new notation made necessary in the performance of the fundamental operations, as $8xA$ ($= 8xy$), $1A_3$ ($= y^2$), and several others.

Cardan, over whose name a shadow has been cast by his selfishness in his intercourse with Tartaglia, is still deserving of credit, particularly for his approximate solution of equations of higher degrees by means of the *regula falsi* which he calls *regula aurea*. Vieta went farther in this direction and evolved a method of approximating the solution of algebraic equations of any degree whatsoever, the method improved by Newton and commonly ascribed to him. Reymers and Bürgi also contributed to these methods of approximation, using the *regula falsi*. We can therefore say that by the beginning of the seventeenth century there were practical methods at hand for calculating the positive real roots of algebraic equations to any desired degree of exactness.

The real theory of algebraic equations is especially due to Vieta. He understood (admitting only positive roots) the relation of the coefficients of equations of the second and third degree to their roots, and also made the surprising discovery that a certain equation of the forty-fifth degree, which had arisen in trig-

* Cantor, II., p. 392.

onometric work, possessed twenty-three roots (in this enumeration he neglected the negative sine). In German writings there are also found isolated statements concerning the analytic theory of equations; for example, Bürgi recognized the connection of a change of sign with a root of the equation. However unimportant these first approaches to modern theories may appear, they prepared the way for ideas which became dominant in later times.

D. THIRD PERIOD.

FROM THE MIDDLE OF THE SEVENTEENTH CENTURY TO THE PRESENT TIME.

The founding of academies and of royal societies characterizes the opening of this period, and is the external sign of an increasing activity in the field of mathematical sciences. The oldest learned society, the Accademia dei Lincei, was organized upon the suggestion of a Roman gentleman, the Duke of Cesi, as early as 1603, and numbered, among other famous members, Galileo. The Royal Society of London was founded in 1660, the Paris Academy in 1666, and the Academy of Berlin in 1700.*

With the progressive development of pure mathematics the contrast between arithmetic, which has to do with discrete quantities, and algebra, which relates rather to continuous quantities, grew more and more

* Cantor, III., pp. 7, 29.

marked. Investigations in algebra as well as in the theory of numbers attained in the course of time great proportions.

The mighty impulse which Vieta's investigations had given influenced particularly the works of Harriot. Building upon Vieta's discoveries, he gave in his *Artis analyticae praxis*, published posthumously in the year 1631, a theory of equations, in which the system of notation was also materially improved. The signs $>$ and $<$ for "greater than" and "less than" originated with Harriot, and he always wrote x^2 for xx and x^3 for xxx, etc. The sign $\times$ for "times" is found almost simultaneously in both Harriot and Oughtred, though due to the latter; Descartes used a period to indicate multiplication, while Leibnitz in 1686 indicated multiplication by $\frown$ and division by $\smile$, although already in the writings of the Arabs the quotient of a divided by b had appeared in the forms $a - b$, a/b, or $\frac{a}{b}$. The form $a : b$ is used for the first time by Clairaut in a work which was published posthumously in the year 1760. Wallis made use in 1655 of the sign ∞ to indicate infinity. Descartes made extensive use of the the form a^n (for positive integral exponents). Wallis explained the expressions x^{-n} and $x^{\frac{1}{n}}$ as indicating the same thing as $1 : x^n$ and $\sqrt[n]{x}$ respectively; but Leibnitz and Newton were the first to recognize the great importance of, and to suggest, a consistent system of notation.

The powers of a binomial engaged the attention
of Pascal in his correspondence with Fermat in 1654,*
which contains the "arithmetic triangle," although,
in its essential nature at least, it had been suggested
by Stifel more than a hundred years before.   This
arithmetic triangle is a table of binomial coefficients
arranged in the following form:

1	1	1	1	1	1	.	.
1	2	3	4	5	6	.	.
1	3	6	10	15	21	.	.
1	4	10	20	35	56	.	.
1	5	15	35	70	126	.	.
1	6	21	56	126	252	.	.
.	.	.	.	.	.	.	.
.	.	.	.	.	.	.	.

so that the nth diagonal, extending upwards from left
to right contains the coefficients of the expansion of
$(a + b)^n$. Pascal used this table for developing figurate
numbers and the combinations of a given number of
elements.   Newton generalized the binomial formula
in 1669, Vandermonde gave an elementary proof in
1764, and Euler in 1770 in his *Anleitung zur Algebra*
gave a proof for any desired exponent.

A series of interesting investigations, for the most
part belonging to the second half of the nineteenth
century, relates to the nature of number and the extension of the number-concept. While among the ancients a "number" meant one of the series of natural

* Cantor, II., p. 684.

numbers only, in the course of time the fundamental operations of arithmetic have been extended from whole to fractional, from positive to negative, from rational and real to irrational and imaginary numbers.

For the addition of natural, or integral absolute, numbers, which by Newton and Cauchy are often termed merely "numbers," the associative and commutative laws hold true, that is,

$$a+b+c=a+(b+c),\ a+b+c=a+c+b.$$

Their multiplication obeys the associative, commutative, and distributive laws, so that

$$abc=(ab)c;\ ab=ba;\ (a+b)c=ac+bc.$$

To these direct operations correspond, as inverses, subtraction and division. The application of these operations to all natural numbers necessitates the introduction of the zero and of negative and fractional numbers, thus forming the great domain of rational numbers, within which these operations are always valid, if we except the one case of division by zero.

This extension of the number-system showed itself in the sixteenth century in the introduction of negative quantities. Vieta distinguished affirmative (positive) and negative quantities. But Descartes was the first to venture, in his geometry, to use the same letter for both positive and negative quantities.

The irrational had been incorporated by Euclid into the mathematical system upon a geometric basis, this plan being followed for many centuries. Indeed

it was not until the most modern times* that a purely
arithmetic theory of irrational numbers was produced
through the researches of Weierstrass, Dedekind, G.
Cantor, and Heine.

Weierstrass proceeds† from the concept of the
whole number. A numerical quantity consists of a
series of objects of the same kind; a number is therefore nothing more than the "combined representation
of one and one and one, etc."‡ By means of subtraction and division we arrive at negative and fractional
numbers. Among the latter there are certain numbers
which, if referred to one particular system, for example to our decimal system, consist of an infinite number of elements, but by transformation can be made
equal to others arising from the combination of a finite
number of elements (e. g., $0.1333\ldots = \frac{2}{15}$). These
numbers are capable of still another interpretation.
But it can be proved that every number formed from
an infinite number of elements of a known species,
and which contains a known finite number of those
elements, possesses a very definite meaning, whether
it is capable of actual expression or not. When a
number of this kind can only be represented by the
infinite number of its elements, and in no other way,
it is an irrational number.

Dedekind§ arranges all positive and negative, in-

*Stolz, *Vorlesungen über allgemeine Arithmetik*, 1885-1886.
† Kossak, *Die Elemente der Arithmetik*, 1872.
‡ Rösler, *Die neueren Definitionsformen der irrationalen Zahlen*, 1886.
§ Dedekind. *Stetigkeit und irrationale Zahlen*, 1872.

tegral and fractional numbers, according to their magnitude, in a system or in a body of numbers (*Zahlen-
körper*), R. A given number, a, divides this system
into the two classes, A_1 and A_2, each containing infinitely many numbers, so that every number in A_1 is
less than every number in A_2. Then a is either the
greatest number in A_1 or the least in A_2. These rational numbers can be put into a one-to-one correspondence with the points of a straight line. It is
then evident that this straight line contains an infinite
number of other points than those which correspond
to rational numbers, that is, the system of rational
numbers does not possess the same continuity as the
straight line, a continuity possible only by the introduction of new numbers. According to Dedekind the
essence of continuity is contained in the following
axiom : "If all the points of a straight line are divided
into two classes such that every point of the first class
lies to the left of every point of the second, then there
exists one point and only one which effects this division of all points into two classes, this separation of
the straight line into two parts." With this assumption it becomes possible to create irrational numbers.
A rational number, a, produces a *Schnitt* or section
$(A_1|A_2)$, with respect to A_1 and A_2, with the characteristic property that there is in A_1 a greatest, or in
A_2 a least number, a. To every one of the infinitely
many points of the straight line which are not covered
by rational numbers, or in which the straight line is

not cut by a rational number, there corresponds one and only one section $(A_1|A_2)$, and each one of these sections defines one and only one irrational number a.

In consequence of these distinctions "the system R constitutes an organized domain of all real numbers of one dimension; by this no more is meant to be said than that the following laws govern : *

I. If $a > \beta$, and $\beta > \gamma$, then a is also $> \gamma$; that is, the number β lies between the numbers a, γ.

II. If a, γ are two distinct numbers, then there are infinitely many distinct numbers β which lie between a and γ.

III. If a is a definite number, then all numbers of the system R fall into two classes, A_1 and A_2, each of which contains infinitely many distinct numbers; the first class A_1 contains all numbers a_1 which are $< a$; the second class A_2 contains all numbers a_2 which are $> a$; the number a itself can be assigned indifferently to either the first or the second class and it is then respectively either the greatest number of the first class, or the least of the second. In every case, the separation of the system R into the two classes A_1 and A_2 is such that every number of the first class A_1 is less than every number of the second class A_2, and we affirm that this separation is effected by the number a.

IV. If the system R of all real numbers is separated into two classes, A_1, A_2, such that every number a_1, of the class A_1 is less than every number a_2 of the class A_1, then there exists one and only one number a by which this separation is effected (the domain R possesses the property of continuity)."

According to the assertion of J. Tannery[†] the fundamental ideas of Dedekind's theory had already appeared in J. Bertrand's text-books of arithmetic and algebra, a statement denied by Dedekind.[‡]

* Dedekind, *Stetigkeit und irrationale Zahlen*, 1872.

† Stolz, *Vorlesungen über allgemeine Arithmetik*, 1885-1886.

‡ Dedekind, *Was sind und was sollen die Zahlen?* 1888.

G. Cantor and Heine* introduce irrational numbers through the concept of a fundamental series. Such a series consists of infinitely many rational numbers, $a_1, a_2, a_3, \ldots a_{n+r}, \ldots$, and it possesses the property that for an assumed positive number ϵ, however small, there is an index n, so that for $n \geq m_1$ the absolute value of the difference between the term a_n and any following term is smaller than ϵ (condition of the convergency of the series of the a's). Any two fundamental series can be compared with each other to determine whether they are equal or which is the greater or the less; they thus acquire the definiteness of a number in the ordinary sense. A number defined by a fundamental series is called a "series number." A series number is either identical with a rational number, or not identical; in the latter case it defines an irrational number. The domain of series numbers consists of the totality of all rational and irrational numbers, that is to say, of all real numbers, and of these only. In this case the domain of real numbers can be associated with a straight line, as G. Cantor has shown.

The extension of the number-domain by the addition of imaginary quantities is closely connected with the solution of equations, especially those of the third degree. The Italian algebraists of the sixteenth century called them "impossible numbers." As proper solutions of an equation, imaginary quantities first

---

* Rösler, *Die neueren Definitionsformen der irrationalen Zahlen*, 1886.

appear in the writings of Albert Girard* (1629). The
expressions "real" and "imaginary" as characteristic
terms for the difference in nature of the roots of an
equation are due to Descartes.† De Moivre and Lambert introduced imaginary quantities into trigonometry, the former by means of his famous proposition
concerning the power $(\cos\phi + i\sin\phi)^n$, first given in
its present form by Euler.‡

Gauss§ added to his great fame by explaining the
nature of imaginary quantities. He brought into gen-
eral use the sign i for $\sqrt{-1}$ first suggested by Euler; ||
he calls $a + bi$ a complex number with the norm
$a^2 + b^2$. The term "modulus" for the quantity $\sqrt{a^2 + b^2}$
comes from Argand (1814), the term "reduced form"
for $r(\cos\phi + i\sin\phi)$, which equals $a + bi$, is due to
Cauchy, and the name "direction coefficient" for the
factor $\cos\phi + i\sin\phi$ first appeared in print in an essay
of Hankel's (1861), although it was in use somewhat
earlier. Gauss, to whom in 1799 it seemed simply
advisable to retain complex numbers,¶ by his explanations in the advertisement to the second treatise on
biquadratic **resi**dues gained for them a triumphant
introduction into arithmetic operations.

The way for the geometric representation of complex quantities was prepared by the observations of

*Cantor, II., p. 718.   †Cantor, II., p. 724.   ‡Cantor, III., p. 68.

§ Hankel, *Die komplexen Zahlen*, 1867, p. 71.

| Beman, "Euler's Use of i to Represent an Imaginary," *Bull. Amer.
Math. Soc.*, March, 1898, p. 274.

¶ Treutlein.

various mathematicians of the seventeenth and eighteenth centuries, among them especially Wallis,* who in solving geometric problems algebraically became aware of the fact that when certain assumptions give two real solutions to a problem as points of a straight line, other assumptions give two "impossible" roots as the points of a straight line perpendicular to the first one. The first satisfactory representation of complex quantities in a plane was devised by Caspar Wessel in 1797, without attracting the attention it deserved. A similar treatment, but wholly independent, was given by Argand in 1806.† But his publication was not appreciated even in France. In the year 1813 there appeared in *Gergonne's Annales* by an artillery officer Français in the city of Metz the outlines of a theory of imaginary quantities the main ideas of which can be traced back to Argand. Although Argand improved his theory by his later work, yet it did not gain recognition until Cauchy entered the lists as its champion. It was, however, Gauss who (1831), by means of his great reputation, made the representation of imaginary quantities in the "Gaussian plane" the common property of all mathematicians.‡

Gauss and Dirichlet introduced general complex numbers into arithmetic. The primary investigations

* Hankel, *Die komplexen Zahlen.* 1867, p. 81.

† Hankel, *Die komplexen Zahlen.* 1867, p. 82.

‡ For a *résumé* of the history of the geometric representation of the imaginary, see Beman, "A Chapter in the History of Mathematics," *Proc. Amer. Assn. Adv. Science,* 1897, pp. 33-50

of Dirichlet in regard to complex numbers, which, together with indications of the proof, are contained in
the *Berichte der Berliner Akademie* for 1841, 1842, and
1846, received material amplifications through Eisenstein, Kummer, and Dedekind.   Gauss, in the development of the real theory of biquadratic residues,
introduced complex numbers of the form $a + bi$, and
Lejeune Dirichlet introduced into the new theory of
complex numbers the notions of prime numbers,
congruences, residue-theorems, reciprocity, etc , the
propositions, however, showing greater complexity
and variety and offering greater difficulties in the way
of proof.*   Instead of the equation $x^4 - 1 = 0$, which
gives as roots the Gaussian units, $+1, -1, +i, -i$,
Eisenstein made use of the equation $x^3 - 1 = 0$ and
considered the complex numbers $a + b\rho$ (ρ being a
complex cube root of unity) the theory resembling that
of the Gaussian numbers $a + bi$, but yet possessing
certain marked differences.   Kummer generalized the
theory still further, using the equation $x^n - 1 = 0$ as
the basis, so that numbers of the form

$$a = a_1 A_1 + a_2 A_2 + a_3 A_3 + \ldots .$$

arise where the a's are real integers and the A's are
roots of the equation $x^n - 1 = 0$.   Kummer also set
forth the concept of ideal numbers, that is, of such
numbers as are factors of prime numbers and possess
the property that there is always a power of these ideal
numbers which gives a real number.   For example,

*Cayley, *Address to the British Association*, etc., 1883.

there exists for the prime number p no rational factorization so that $p^3 = A \cdot B$ (where A is different from p and p^2); but in the theory of numbers formed from the twenty-third roots of unity there are prime numbers p which satisfy the condition named above. In this case p is the product of two ideal numbers, of which the third powers are the real numbers A and B, so that $p^3 = A \cdot B$. In the later development given by Dedekind the units are the roots of any irreducible equation with integral numerical coefficients. In the case of the equation $x^2 - x + 1 = 0$, $\frac{1}{2}(1 + i\sqrt{3})$, that is to say, the ρ of Eisenstein, is to be regarded as integral.

In tracing out the nature of complex numbers, H. Grassmann, Hamilton, and Scheffler have arrived at peculiar discoveries. Grassmann, who also materially developed the theory of determinants, investigated in his treatise on directional calculus (*Ausdehnungslehre*) the addition and multiplication of complex numbers. In like manner, Hamilton originated the calculus of quaternions, a method of calculation regarded with especial favor in England and America and justified by its relatively simple applicability to spherics, to the theory of curvature, and to mechanics.

The complete double title* of H. Grassmann's chief work which appeared in the year 1844, as translated, is: "The Science of Extensive Quantities or Directional Calculus (*Ausdehnungslehre*). A New

---

* V. Schlegel, *Grassmann, sein Leben und seine Werke.*

Mathematical Theory, Set Forth and Elucidated by
Applications.   Part First, Containing the Theory of
Lineal Directional Calculus.   The Theory of Lineal
Directional Calculus, A New Branch of Mathematics,
Set Forth and Elucidated by Applications to the
Remaining Branches of Mathematics, as well as to
Statics, Mechanics, the Theory of Magnetism and
Crystallography."   The favorable criticisms of this
wonderful work by Gauss, who discovered that "the
tendencies of the book partly coincided with the paths
upon which he had himself been travelling for half a
century," by Grunert, and by Möbius who recognised
in Grassmann "a congenial spirit with respect to
mathematics, though not to philosophy," and who
congratulated Grassmann upon his "excellent work,"
were not able to secure for it a large circle of readers.
As late as 1853 Möbius stated that "Bretschneider
was the only mathematician in Gotha who had assured
him that he had read the *Ausdehnungslehre* through."

Grassmann received the suggestion for his researches from geometry, where A, B, C, being points
of a straight line, $AB + BC = AC$.*   With this he
combined the propositions which regard the parallelo
gram as the product of two adjacent sides, thus introducing new products for which the ordinary rules of
multiplication hold so long as there is no permutation
of factors, this latter case requiring the change of

---

*Grassmann, *Die Ausdehnungslehre von 1844 oder die lineale Ausdeh-
nungslehre, ein neuer Zweig der Mathematik*.   Zweite Auflage, 1878.

signs. More exhaustive researches led Grassmann to regard as the sum of several points their center of gravity, as the product of two points the finite line-segment between them, as the product of three points the area of their triangle, and as the product of four points the volume of their pyramid. Through the study of the *Barycentrischer Calcül* of Möbius, Grassmann was led still further. The product of two line-segments which form a parallelogram was called the "external product" (the factors can be permuted only by a change of sign), the product of one line-segment and the perpendicular projection of another upon it formed the "internal product" (the factors can here be permuted without change of sign). The introduction of the exponential quantity led to the enlargement of the system, of which Grassmann permitted a brief survey to appear in *Grunert's Archiv* (1845).[*]

Hamilton[†] gave for the first time, in a communication to the Academy of Dublin in 1844, the values i, j, k, so characteristic of his theory. The *Lectures on Quaternions* appeared in 1853, the *Elements of Quaternions* in 1866. From a fixed point O let a line[‡] be drawn to the point P having the rectangular co-ordinates x, y, z. Now if i, j, k represent fixed coefficients (unit distances on the axes), then

[*] Translated by Beman, *Analyst*, 1881, pp. 96, 114.

[†] Unverzagt, *Theorie der goniometrischen und longimetrischen Quaternionen*, 1876.

[‡] Cayley, A.. "On Multiple Algebra," in *Quarterly Journal of Mathematics*, 1887.

$$V = ix + jy + kz$$

is a vector, and this additively joined to the "pure quantity" or "scalar" w produces the quaternion

$$Q = w + ix + jy + kz.$$

The addition of two quaternions follows from the usual formula

$$Q + Q' = w + w' + i(x + x') + j(y + y') + k(z + z')$$

But in the case of multiplication we must place

$$i^2 = j^2 = k^2 = -1, \quad i = jk = -kj, \quad j = ki = -ik,$$
$$k = ij = -ji,$$

so that we obtain

$$Q \cdot Q' = ww' - xx' - yy' - zz'$$
$$+ i(wx' + xw' + yz' - zy')$$
$$+ j(wy' + yw' + zx' - xz')$$
$$+ k(wz' + zw' + xy' - yx')$$

On this same subject Scheffler published in 1846 his first work, *Ueber die Verhältnisse der Arithmetik zur Geometrie*, in 1852 the *Situationscalcul*, and in 1880 the *Polydimensionalen Grössen*. For him[*] the vector r in three dimensions is represented by

$$r = a \cdot e^{a\sqrt{-1}} \cdot e^{\beta\sqrt{+1}}, \text{ or}$$
$$r = x + y\sqrt{-1} + z\sqrt{-1} \cdot \sqrt{+1}, \text{ or}$$

$r = x + y \cdot i + z \cdot i \cdot i_1$ where $i = \sqrt{-1}$ and $i_1 = \sqrt{+1}$ are turning factors of an angle of $90°$ in the plane of xy and xz. In Scheffler's theory the distributive law does not always hold true for multiplication, that is to say, $a(b + c)$ is not always equivalent to $ab + ac$.

Investigations as to the extent of the domain in

[*] Unverzagt, *Ueber die Grundlagen der Rechnung mit Quaternionen*, 1881.

which with certain assumptions the laws of the elementary operations of arithmetic are valid have led to the establishment of a calculus of logic.[*] To this class of investigations there belong, besides Grassmann's *Formenlehre* (1872), notes by Cayley and Ellis, and in particular the works of Boole, Schröder, and Charles Peirce.

A minor portion of the modern theory of numbers or higher arithmetic, which concerns the theories of congruences and of forms, is made up of continued fractions. The algorism leading to the formation of such fractions, which is also used in calculating the greatest common measure of two numbers, reaches back to the time of Euclid. The combination of the partial quotients in a continued fraction originated with Cataldi,[†] who in the year 1613 approximated the value of square roots by this method, but failed to examine closely the properties of the new fractions.

Daniel Schwenter was the first to make any material contribution (1625) towards determining the convergents of continued fractions. He devoted his attention to the reduction of fractions involving large numbers, and determined the rules now in use for calculating the successive convergents. Huygens and Wallis also labored in this field, the latter discovering the general rule, together with a demonstration, which combines the terms of the convergents

[*] Schröder, *Der Operationskreis des Logikcalculs*, 1877.

[†] Cantor, II., p. 695.

$$\frac{p_n}{q_n}, \quad \frac{p_{n-1}}{q_{n-1}}, \quad \frac{p_{n-2}}{q_{n-2}}$$

in the following manner:

$$\frac{p_n}{q_n} = \frac{a_n p_{n-1} + b_n p_{n-2}}{a_n q_{n-1} + b_n q_{n-2}}.$$

The theory of continued fractions received its greatest development in the eighteenth century with Euler,[*] who introduced the name *fractio continua* (the German term *Kettenbruch* has been used only since the beginning of the nineteenth century). He devoted his attention chiefly to the reduction of continued fractions to the form of infinite products and series, and doubtless in this way was led to the attempt to give the convergents independent form, that is to discover a general law by means of which it would be possible to calculate any required convergent without first obtaining the preceding ones. Although Euler did not succeed in discovering such a law, he created an algorism of some value. This, however, did not bring him essentially nearer the goal because, in spite of the example of Cramer, he neglected to make use of determinants and thus to identify himself the more closely with the pure theory of combinations. From this latter point of view the problem was attacked by Hindenburg and his pupils Burckhardt and Rothe. Still, those who proceed from the theory of combinations alone know continued fractions only from one side; the method of independent presentation allows

* Cantor, III., p. 670.

the calculation of the desired convergent from both sides, forward as well as backward, to the practical value of which Dirichlet has testified.

Only in modern times has the calculus of determinants been employed in this field, together with a combinatory symbol, and the first impulse in this direction dates from the Danish mathematician Ramus (1855). Similar investigations were begun, however, by Heine, Möbius, and S. Günther, leading to the formation of "continued fractional determinants." The irrationality of certain infinite continued fractions * had been investigated before this by Legendre, who, like Gauss, gave the quotient of two power series in the form of a continued fraction. By means of the application of continued fractions it can be shown that the quantities e^x (for rational values of x), e, π, and π^2 cannot be rational (Lambert, Legendre, Stern). It was not until very recent times that the transcendental nature of e was established by Hermite, and that of π by F. Lindemann.

In the theory of numbers strictly speaking, quite difficult problems concerning the properties of numbers were solved by the first exponents of that study, Euclid and Diophantus. Any considerable advance was impossible, however, as long as investigations had to be conducted † without an adequate numerical notation, and almost exclusively with the aid of an algebra

* Treutlein.

† Legendre, *Théorie des nombres*, 1st ed. 1798, 3rd ed. 1830.

just developing under the guise of geometry. Until
the time of Vieta and Bachet there is no essential advance to be noted in the theory of numbers. The
former solved many problems in this field, and the
latter gave in his work *Problèmes plaisants et délectables*
a satisfactory treatment of indeterminate equations
of the first degree. Still later the first stones for the
foundation of a theory of numbers were laid by Fermat, who had carefully studied Diophantus and into
whose works as elaborated by Bachet he incorporated
valuable additional propositions. The great mass of
propositions which can be traced back to him he gave
for the most part without demonstration, as for example the following statement :

"Every prime number of the form $4n+1$ is the
sum of two squares; a prime number of the form
$8n+1$ has at the same time the three forms y^2+z^2,
y^2+2z^2, y^2-2z^2; every prime number of the form
$8n+3$ appears as y^2+2z^2, every one of the form $8n+7$
appears as y^2-2z^2." Further, "Any number can be
formed by the addition of three cubes, of four squares,
of five fifth powers, etc."

Fermat proved that the area of a Pythagorean
right-angled triangle, for example a triangle with the
sides 3, 4, and 5, cannot be a square. He was also
the first to obtain the solution of the equation ax^2+
$1=y^2$, where a is not a square; at all events, he
brought this problem to the attention of English
mathematicians, among whom Lord Brouncker dis-

covered a solution which found its way into the works of Wallis. Many of Fermat's theorems belong to "the finest propositions of higher mathematics,"* and possess the peculiarity that they can easily be discovered by induction, but that their demonstrations are extremely difficult and yield only to the most searching investigation. It is just this which imparts to higher arithmetic that magic charm which made it a favorite with the early geometers, not to speak of its inexhaustible treasure-house in which it far exceeds all other branches of pure mathematics.

After Fermat, Euler was the first again to attempt any serious investigations in the theory of numbers. To him we owe, among other things, the first scientific solution of the chess board problem, which requires that the knight, starting from a certain square, shall in turn occupy all sixty-four squares, and the further proposition that the sum of four squares multiplied into another similar sum also gives the sum of four squares. He also discovered demonstrations of various propositions of Fermat, as well as the general solution of indeterminate equations of the second degree with two unknowns on the hypothesis that a special solution is known, and he treated a large number of other indeterminate equations, for which he discovered numerous ingenious solutions.

Euler (as well as Krafft) also occupied himself

* Gauss, *Werke*, II., p. 152.

with amicable numbers.*   These numbers, which are
mentioned by Iamblichus as being known to the
Pythagoreans, and which are mentioned by the Arab
Tabit ibn Kurra, suggested to Descartes the discovery
of a law of formation, which is given again by Van
Schooten.   Euler made additions to this law and deduced from it the proposition that two amicable num
bers must possess the same number of prime factors.
The formation of amicable numbers depends either
upon the solution of the equation $xy + ax + by + c = 0$,
or upon the factoring of the quadratic form $ax^2 + bxy$
$+ cy^2$.

Following Euler, Lagrange was able to publish
many interesting results in the theory of numbers.
He showed that any number can be represented as
the sum of four or less squares, and that a real root
of an algebraic equation of any degree can be con-
verted into a continued fraction.   He was also the
first to prove that the equation $x^2 - Ay^2 = 1$ is always
soluble in integers, and he discovered a general method
for the derivation of propositions concerning prime
numbers.

Now the development of the theory of numbers
bounds forward in two mighty leaps to Legendre and
Gauss.   The valuable treatise of the former, *Essai sur
la théorie des nombres*, which appeared but a few years
before Gauss's *Disquisitiones arithmeticae*, contains an
epitome of all results that had been published up to

* Seelhoff, "Befreundete Zahlen." *Hoppe Arch.*, Bd. 70.

that time, besides certain original investigations, the
most brilliant being the law of quadratic reciprocity,
or, as Gauss called it, the *Theorema fundamentale in
doctrina de residuis quadratis.* This law gives a relationship between two odd and unequal prime numbers
and can be enunciated in the following words:

"Let $\left(\dfrac{m}{n}\right)$ be the remainder which is left after divid-

ing $m^{\frac{n-1}{2}}$ by n, and let $\left(\dfrac{n}{m}\right)$ be the remainder left after

dividing $n^{\frac{m-1}{2}}$ by m. These remainders are always
$+1$ or -1. Whatever then the prime numbers m
and n may be, we always obtain $\left(\dfrac{n}{m}\right) = \left(\dfrac{m}{n}\right)$ in case the
numbers are not both of the form $4x+3$. But if both
are of the form $4x+3$, then we have $\left(\dfrac{n}{m}\right) = -\left(\dfrac{m}{n}\right)$."

These two cases are contained in the formula

$$\left(\frac{n}{m}\right) = (-1)^{\frac{m-1}{2} \cdot \frac{n-1}{2}} \cdot \left(\frac{m}{n}\right).$$

Bachet having exhausted the theory of the indeterminate equation of the first degree with two unknowns,
an equation which in Gauss's notation appears in the
form $x \equiv a \pmod{b}$, identical with $\dfrac{x}{b} = y+a$, mathematicians began the study of the congruence $x^2 \equiv m$
$\pmod{n}$. Fermat was aware of a few special cases of
the complete solution; he knew under what conditions
± 1, 2, ± 3, 5 are quadratic residues or non-residues
of the odd prime number m.* For the cases -1 and

* Baumgart, "Ueber das quadratische Reciprocitätsgesetz," in *Schlömilch's Zeitschrift,* Bd. 30, III. Abt.

± 3 the demonstrations originate with Euler, for ± 2 and ± 5 with Lagrange. It was Euler, too, who gave the propositions which embrace the law of quadratic reciprocity in the most general terms, although he did not offer a complete demonstration of it. The famous demonstration of Legendre (in *Essai sur la théorie des nombres*, 1798) is also, as yet, incomplete. In the year 1796 Gauss submitted, without knowing of Euler's work, the first unquestionable demonstration—a demonstration which possesses at the same time the peculiarity that it embraces the principles which were used later. In the course of time Gauss adduced no less than eight proofs for this important law, of which the sixth (chronologically the last) was simplified almost simultaneously by Cauchy, Jacobi, and Eisenstein. Eisenstein demonstrated in partic ular that the quadratic, the cubic and the biquadratic laws are all derived from a common source. In the year 1861 Kummer worked out with the aid of the theory of forms two demonstrations for the law of quadratic reciprocity, which were capable of gene- ralization for the nth-power residue. Up to 1890 twenty-five distinct demonstrations of the law of quadratic reciprocity had been published; they make use of induction and reduction, of the partition of the perigon, of the theory of functions, and of the theory of forms. In addition to the eight demonstrations by Gauss which have already been mentioned, there are four by Eisenstein, two by Kummer, and one each by Jacobi, Cauchy, Liouville, Lebesgue, Genocchi, Stern, Zeller, Kronecker, Bouniakowsky, Schering, Petersen, Voigt, Busche, and Pepin.

However much is due to the co-operation of mathematicians of different periods, yet to Gauss unquestionably belongs the merit of having contributed in his *Disquisitiones arithmeticae* of 1801 the most important part of the elementary development of the theory of numbers. Later investigations in this branch have their root in the soil which Gauss prepared. Of such investigations, which were not pursued until after the introduction of the theory of elliptic transcendents, may be mentioned the propositions of Jacobi in regard to the number of decompositions of a number into two, four, six, and eight squares,* as well as the investigations of Dirichlet in regard to the equation

$$x^n + y^n = z^n.$$

His work in the theory of numbers was Dirichlet's favorite pursuit.† He was the first to deliver lectures on the theory of numbers in a German university and was able to boast of having made the *Disquisitiones arithmeticae* of Gauss transparent and intelligible—a task in which a Legendre, according to his own avowal, was unsuccessful.

Dirichlet's earliest treatise, *Mémoire sur l'impossibilité de quelques équations indéterminés du cinquième degré* (submitted to the French Academy in 1825),

* Dirichlet, "Gedächtnisrede auf Jacobi," *Crelle's Journal*, Bd. 52.
† Kummer, "Gedächtnisrede auf Lejeune-Dirichlet," in *Berl. Abh.* 1860.

deals with the proposition, stated by Fermat without
demonstration, that "the sum of two powers having
the same exponent can never be equal to a power of
the same exponent, when these powers are of a degree
higher than the second." Euler and Legendre had
proved this proposition for the third and fourth powers; Dirichlet discusses the sum of two fifth powers
and proves that for integral numbers $x^5 + y^5$ cannot
be equal to az^5. The importance of this work lies in
its intimate relationship to the theory of forms of
higher degree. Dirichlet's further contributions in the
field of the theory of numbers contain elegant demonstrations of certain propositions of Gauss in regard
to biquadratic residues and the law of reciprocity,
which were published in 1825 in the Göttingen *Ge-
lehrte Anzeigen*, as well as with the determination of
the class-number of the quadratic form for any given
determinant. His "applications of analysis to the
theory of numbers are as noteworthy in their way as
Descartes's applications of analysis to geometry. They
would also, like the analytic geometry, be recognized
as a new mathematical discipline if they had been extended not to certain portions only of the theory of
number, but to all its problems uniformly.[*]

The numerous investigations into the properties
and laws of numbers had led in the seventeenth century[†] to the study of numbers in regard to their divis-

[*] Kummer, "Gedächtnisrede auf Lejeune-Dirichlet." *Berl. Abh.* 1860.
[†] Seelhoff. "Geschichte der Faktorentafeln," in *Hoppe Arch.*, Bd. 70.

ors. For almost two thousand years Eratosthenes's
"sieve" remained the only method of determining
prime numbers. In the year 1657 Franz van Schooten
published a table of prime numbers up to ten thousand. Eleven years later Pell constructed a table of
the least prime factors (with the exception of 2 and 5)
of all numbers up to 100 000. In Germany these
tables remained almost unknown, and in the year
1728 Poëtius published independently a table of factors for numbers up to 100 000, an example which
was repeatedly imitated. Krüger's table of 1746 includes numbers up to 100 000; that of Lambert of
1770, which is the first to show the arrangement
used in more modern tables, includes numbers up to
102 000. Of the six tables which were prepared between the years 1770 and 1811 that of Felkel is interesting because of its singular fate; its publication by
the *Kaiserlich königliches Aerarium* in Vienna was
completed as far as 408 000; the remainder of the
manuscript was then withheld and the portion already
printed was used for manufacturing cartridges for the
last Turkish war of the eighteenth century. In the
year 1817 there appeared in Paris Burckhardt's *Table
des diviseurs pour tous les nombres du 1er, 2^e, 3^e million.*
Between 1840 and 1850 Crelle communicated to the
Berlin Academy tables of factors for the fourth, fifth,
and sixth million, which, however, were not published. Dase, who is known for his arithmetic genius, was to make the calculations for the seventh to

the tenth million, having been designated for that
work by Gauss, but he died in 1861 before its completion.   Since 1877 the British Association has been
having these tables continued by Glaisher with the
assistance of two computers. The publication of
tables of factors for the fourth million was completed
in 1879.

In the year 1856 K. G. Reuschle published his
tables for use in the theory of numbers, having been
encouraged to undertake the work by his correspondence with Jacobi. They contain the resolution of
numbers of the form $10^n - 1$ into prime factors, up to
$n = 242$, and numerous similar results for numbers of
the form $a^n - 1$, and a table of the resolution of prime
numbers $p = 6n + 1$ into the forms

$$p = A^2 + 3B^2 \text{ and } 4p = C^2 + 27M^2,$$

as they occur in the treatment of cubic residues and
in the partition of the perigon.

Of greatest importance for the advance of the science of algebra as well as that of geometry was the
development of the theories of symmetric functions,
of elimination, and of invariants of algebraic forms,
as they were perfected through the application of projective geometry to the theory of equations.[*]

The first formulas for calculating symmetric functions (sums of powers) of the roots of an algebraic
equation in terms of its coefficients are due to Newton.

---

[*] A. Brill. *Antrittsrede in Tübingen*, 1884. Manuscript.

Waring also worked in this field (1770) and developed
a theorem, which Gauss independently discovered
(1816), by means of which any symmetric function
may be expressed in terms of the elementary symmetric functions.   This is accomplished directly by a
method devised by Cayley and Sylvester, through laws
due to the former in regard to the weight of sym
metric functions.   The oldest tables of symmetric
functions (extending to the tenth degree) were published by Meyer-Hirsch in his collection of problems
(1809). The calculation of these functions, which was
very tedious, was essentially simplified by Cayley and
Brioschi.

The resultant of two equations with one unknown,
or, what is the same, of two forms with two homogeneous variables, was given by Euler (1748) and by
Bézout (1764).   To both belongs the merit of having
reduced the determination of the resultant to the solution of a system of linear equations.* Bézout introduced the name "resultant" (De Morgan suggested
"eliminant") and determined the degree of this function.   Lagrange and Poisson also investigated questions of elimination ; the former stated the condition
for common multiple-roots; the latter furnished a
method of forming symmetric functions of the common values of the roots of a system of equations. The
further advancement of the theory of elimination was
made by Jacobi, Hesse, Sylvester, Cayley, Cauchy,

* Salmon, *Higher Algebra*

Brioschi, and Gordan. Jacobi's memoir,[*] which represented the resultant as a determinant, threw light
at the same time on the aggregate of coefficients belonging to the resultant and on the equations in which
the resultant and its product by another partially arbitrary function are represented as functions of the
two given forms. This notion of Jacobi gave Hesse
the impulse to pursue numerous important investigations, especially on the resultant of two equations,
which he again developed in 1843 after Sylvester's
dialytic method (1840); then in 1844, "on the elimination of the variables from three algebraic equations
with two variables"; and shortly after "on the points
of inflexion of plane curves." Hesse placed the main
value of these investigations, not in the form of the
final equation, but in the insight into the composition
of the same from known functions. Thus he came
upon the functional determinant of three quadratic
prime forms, and further upon the determinant of the
second partial differential coefficients of the cubic
form, and upon its Hessian determinant, whose geometric interpretation furnished the interesting result
that in the general case the points of inflexion of a
plane curve of the nth order are given by its complete
intersection with a curve of order $3(n-2)$. This result was previously known for curves of the third
order, having been discovered by Plücker. To Hesse
is further due the first important example of the re-

*O. H. Noether. *Schlömilch's Zeitschrift*, Bd. 20.

moval of factors from resultants, in so far as these factors are foreign to the real problem to be solved. Hesse, always extending the theory of elimination, in 1849 succeeded in producing, free from all superfluous factors, the long-sought equation of the 14th degree upon which the double tangents of a curve of the 4th order depend.

The method of elimination used by Hesse* in 1843 is the dialytic method published by Sylvester in 1840; it gives the resultant of two functions of the mth and nth orders as a determinant, in which the coefficients of the first enter into n rows, and those of the second into m rows. It was Sylvester also, who in 1851 introduced the name "discriminant" for the function which expresses the condition for the existence of two equal roots of an algebraic equation; up to this time, it was customary, after the example of Gauss, to say "determinant of the function."

The notion of invariance, so important for all branches of mathematics to-day, dates back in its beginnings to Lagrange†, who in 1773 remarked that the discriminant of the quadratic form $ax^2 + 2bxy + cy^2$ remains unaltered by the substitution of $x + \lambda y$ for x. This unchangeability of the discriminant by linear transformation, for binary and ternary quadratic forms, was completely proved by Gauss (1801); but that the discriminant in general and in every case remains invariant by linear transformation,

* Matthiessen, p. 99.　　　† Salmon, *Higher Algebra*.

G. Boole (1841) recognized and first demonstrated.
In 1845, Cayley, adding to the treatment of Boole,
found that there are still other functions which possess
invariant properties in linear transformation, showed
how to determine such functions and named them
"hyperdeterminants." This discovery of Cayley developed rapidly into the important theory of invariants, particularly through the writings of Cayley,
Aronhold, Boole, Sylvester, Hermite, and Brioschi,
and then through those of Clebsch, Gordan, and
others. After the appearance of Cayley's first paper,
Aronhold made an important contribution by determining the invariants S and T of a ternary form, and
by developing their relation to the discriminant of
the same form. From 1851 on, there appeared a series of important articles by Cayley and Sylvester.
The latter created in these a large part of the termin
ology of to-day, especially the name "invariant"
(1851). In the year 1854, Hermite discovered his law
of reciprocity, which states that to every covariant or
invariant of degree ρ and order r of a form of the mth
order, corresponds also a covariant or invariant of
degree m and of order r of a form of the ρth order.
Clebsch and Gordan used the abbreviation b_x^n, introduced for binary forms by Aronhold, in their fundamental developments, e. g., in the systematic extension of the process of transvection in forming
invariants and covariants, already known to Cayley
in his preliminary investigations, in the folding-pro-

cess of forming elementary covariants, and in the formation of simultaneous invariants and covariants, in particular the combinants. Gordan's theorem on the finiteness of the form-system constitutes the most important recent advance in this theory; this theorem states that there is only a finite number of invariants and covariants of a binary form or of a system of such forms. Gordan has also given a method for the formation of the complete form-system, and has carried out the same for the case of binary forms of the fifth and sixth orders. Hilbert (1890) showed the finiteness of the complete systems for forms of n variables.[*]

To refer in a word to the great significance of the theory of invariants for other branches of mathematics, let it suffice to mention that the theory of binary forms has been transferred by Clebsch to that of ternary forms (in particular for equations in line co-ordinates) ; that the form of the third order finds its representation in a space-curve of the third order, while binary forms of the fourth order play a great part in the theory of plane curves of the third order, and assist in the solution of the equation of the fourth degree as well as in the transformation of the elliptic integral of the first class into Hermite's normal form ; finally that combinants can be effectively introduced in the transformation of equations of the fifth and sixth degrees. The results of investigations by Clebsch, Weierstrass, Klein, Bianchi, and Burckhardt, have shown the great significance of the theory of invariants for the theory of the hyperelliptic and Abelian functions. This theory has been further used by Christoffel and Lipschitz in the representation of the line-element, by Sylvester, Halphen, and Lie in the case of reciprocants or differential invariants in the theory of dif-

* Meyer, W. F., "Bericht über den gegenwärtigen Stand der Invariantentheorie." *Jahresbericht der deutschen Mathematiker-Vereinigung.* Bd. 1.

ferential equations, and by Beltrami in his differential parameter
in the theory of curvature of surfaces. Irrational invariants also
have been proposed in various articles by Hilbert.

The theory of probabilities assumed form under
the hands of Pascal and Fermat.* In the year 1654,
a gambler, the Chevalier de Méré, had addressed two
inquiries to Pascal as follows: " In how many throws
with dice can one hope to throw a double six," and
"In what ratio should the stakes be divided if the
game is broken up at a given moment?" These two
questions, whose solution was for Pascal very easy,
were the occasion of his laying the foundation of a
new science which was named by him "Géométrie du
hasard." At Pascal's invitation, Fermat also turned
his attention to such questions, using the theory of
combinations. Huygens soon followed the example
of the two French mathematicians, and wrote in 1656†
a small treatise on games of chance. The first to
apply the new theory to economic sciences was the
"grand pensioner" Jean de Witt, the celebrated pupil
of Descartes. He made a report in 1671 on the manner of determining the rate of annuities on the basis
of a table of mortality. Hudde also published investigations on the same subject. "Calculation of
chances" (*Rechnung über den Zufall*) received comprehensive treatment at the hand of Jacob Bernoulli in
his *Ars conjectandi* (1713), printed eight years after the
death of the author, a book which remained forgotten

* Cantor, II., p. 688.   † Cantor, II., p. 692.

until Condorcet called attention to it. Since Bernoulli, there has scarcely been a distinguished algebraist who has not found time for some work in the theory of probabilities.

To the method of least squares Legendre gave the name in a paper on this subject which appeared in 1805.* The first publication by Gauss on the same subject appeared in 1809, although he was in possession of the method as early as 1795. The honor is therefore due to Gauss for the reason that he first set forth the method in its present form and turned it to practical account on a large scale. The apparent inspiration for this investigation was the discovery of the first planetoid Ceres on the first of January, 1801, by Piazzi. Gauss calculated by new methods the orbit of this heavenly body so accurately that the same planetoid could be again found towards the end of the year 1801 near the position given by him. The investigations connected with this calculation appeared in 1809 as *Theoria motus corporum coelestium*, etc. The work contained the determination of the position of a heavenly body for any given time by means of the known orbit, besides the solution of the difficult problem to find the orbit from three observations. In order to make the orbit thus determined agree as closely as possible with that of a greater number of observations, Gauss applied the process

---

*Merriman, M., "List of Writings relating to the Method of Least Squares." *Trans. Conn. Acad.*, Vol. IV.

discovered by him in 1795. The object of this was "so to combine observations which serve the purpose of determining unknown quantities, that the unavoidable errors of observation affect as little as possible the values of the numbers sought." For this purpose Gauss gave the following rule*: "Attribute to each error a moment depending upon its value, multiply the moment of each possible error by its probability and add the products. The error whose moment is equal to this sum will have to be designated as the mean." As the simplest arbitrary function of the error which shall be the moment of the latter, Gauss chose the square. Laplace published in the year 1812 a detailed proof of the correctness of Gauss's method.

Elementary presentations of the theory of combinations are found in the sixteenth century, e. g., by Cardan, but the first great work is due to Pascal. In this he uses his arithmetic triangle, in order to determine the number of combinations of m elements of the nth class. Leibnitz and Jacob Bernoulli produced much new material by their investigations. Towards the end of the eighteenth century, the field was cultivated by a number of German scholars, and there arose under the leadership of Hindenburg the "combinatory school,"† whose followers added to the development of the binomial theorem. Superior to them all in systematic proof is Hindenburg, who separated

---

* Gerhardt, *Geschichte der Mathematik in Deutschland*, 1877.

† Gerhardt, *Geschichte der Mathematik in Deutschland*, 1877.

polynomials into a first class of the form $a + b + c + d + \ldots$ and into a second, $a + bx + cx^2 + dx^3 + \ldots$. He perfected what was already known, and gave the lacking proofs to a number of theorems, thus earning the title of "founder of the theory of combinatory analysis."

The combinatory school, which included Eschenbach, Rothe, and especially Pfaff, in addition to its distinguished founder, produced a varied literature, and commanded respect because of its elegant formal results. But, in its aims, it stood so far outside the domain of the new and fruitful theories cultivated especially by such French mathematicians as Lagrange and Laplace, that it remained without influence in the further development of mathematics, at least at the beginning of the nineteenth century.

In the domain of infinite series,[*] many cases which reduce for the most part to geometric series, were treated by Euclid, and to a greater degree by Apollonius. The Middle Ages added nothing essential, and it remained for more recent generations to make important contributions to this branch of mathematical knowledge. Saint-Vincent and Mercator developed independently the series for $\log(1 + x)$, Gregory those for $\tan^{-1}x$, $\sin x$, $\cos x$, $\sec x$, $\csc x$. In the writings of the latter are also found, in the treatment of infinite series, the expressions "convergent" and "divergent." Leibnitz was led to infinite series, through consideration of finite arithmetic series. He realized at the same time the necessity of examining

* Reiff, R., *Geschichte der unendlichen Reihen*. Tübingen, 1889.

more closely into the convergence and divergence of series. This necessity was also felt by Newton, who used infinite series in a manner similar to that of Apollonius in the solution of algebraic and geometric problems, especially in the determination of areas, and consequently as equivalent to integration.

The new ideas introduced by Leibnitz were further developed by Jacob and John Bernoulli. The former found the sums of series with constant terms, the latter gave a general rule for the development of a function into an infinite series. At this time there were no exact criteria for convergence, except those suggested by Leibnitz for alternating series.

During the years immediately following, essential advances in the formal treatment of infinite series were made. De Moivre wrote on recurrent series and exhausted almost completely their essential properties. Taylor's and Maclaurin's closely related series appeared, Maclaurin developing a rigorous proof of Taylor's theorem, giving numerous applications of it, and stating new formulas of summation. Euler displayed the greatest skill in the handling of infinite series, but troubled himself little about convergence and divergence. He deduced the exponential from the binomial series, and was the first to develop rational functions into series of sines and cosines of integral multiple arguments.* In this manner he defined the coefficients of a trigonometric series by

* Reiff, *Geschichte der unendlichen Reihen*, 1889, pp. 105, 127.

definite integrals without applying these important
formulas to the development of arbitrary functions
into trigonometric series. This was first accomplished
by Fourier (1822), whose investigations were completed by Riemann and Cauchy. The investigation
was brought to a temporary termination by Dirichlet
(1829), in so far as by rigid methods he gave it a scientific foundation and introduced general and complex investigations on the convergence of series.*
From Laplace date the developments into series of
two variables, especially into recurrent series. Legendre furnished a valuable extension of the theory
of series by the introduction of spherical functions.

With Gauss begin more exact methods of treatment in this as in nearly all branches of mathematics,
the establishment of the simplest criteria of convergence, the investigation of the remainder, and the
continuation of series beyond the region of convergence. The introduction to this was the celebrated
series of Gauss:

$$1 + \frac{\alpha \cdot \beta}{1 \cdot \gamma} x + \frac{\alpha(\alpha+1)\beta(\beta+1)}{1 \cdot 2 \cdot \gamma(\gamma+1)} x^2 + \dots,$$

which Euler had already handled but whose great
value he had not appreciated.† The generally accepted naming of this series as "hypergeometric" is
due to J. F. Pfaff, who proposed it for the general
series in which the quotient of any term divided by the

---

* Kummer, "Gedächtnissrede auf Lejeune-Dirichlet." *Berliner Abhandlungen*, 1860.

† Reiff, *Geschichte der unendlichen Reihen*, 1889, p. 161.

preceding is a function of the index. Euler, following Wallis, used the same name for the series in which
that quotient is an integral linear function of the
index.* Gauss, probably influenced by astronomical
applications, stated that his series, by assuming certain special values of a, β, γ, could take the place
of nearly all the series then known; he also investigated the essential properties of the function represented by this series and gave for series in general an
important criterion of convergence. We are indebted
to Abel (1826) for important investigations on the continuity of series.

The idea of uniform convergence arose from the
study of the behavior of series in the neighborhood of
their discontinuities, and was expressed almost simultaneously by Stokes and Seidel (1847–1848). The
latter calls a series uniformly convergent when it represents a discontinuous function of a quantity x, the
separate terms of which are continuous, but in the
vicinity of the discontinuities is of such a nature that
values of x exist for which the series converges as
slowly as desired.†

On account of the lack of immediate appreciation
of Gauss's memoir of 1812, the period of the discovery
of effective criteria of convergence and divergence‡
may be said to begin with Cauchy (1821). His meth-

---

* Riemann, *Werke*, p. 78.

† Reiff, *Geschichte der unendlichen Reihen*, 1889, p. 207.

‡ Pringsheim, "Allgemeine Theorie der Divergenz und Konvergenz von
Reihen mit positiven Gliedern," *Math. Annalen*, XXXV.

ods of investigation, as well as the theorems on infinite series with positive terms published between 1832 and 1851 by Raabe, Duhamel, De Morgan, Bertrand, Bonnet, and Paucker, set forth special criteria, for they compare generally the nth term with particular functions of the form a^n, n^k, $n(\log n)^l$ and others. Criteria of essentially more general nature were first discovered by Kummer (1835), and were generalized by Dini (1867). Dini's researches remained for a time, at least in Germany, completely unknown. Six years later Paul du Bois-Reymond, starting with the same fundamental ideas as Dini, discovered anew the chief results of the Italian mathematician, worked them out more thoroughly and enlarged them essentially to a system of convergence and divergence criteria of the first and second kind, according as the general term of the series a_n or the quotient $a_{n+1} : a_n$ is the basis of investigation. Du Bois-Reymond's results were completed and in part verified somewhat later by A. Pringsheim.

After the solution of the algebraic equations of the third and fourth degrees was accomplished, work on the structure of the system of algebraic equations in general could be undertaken. Tartaglia, Cardan, and Ferrari laid the keystone of the bridge which led from the solution of equations of the second degree to the complete solution of equations of the third and fourth degrees. But centuries elapsed before an Abel threw a flood of light upon the solution of higher equations.

Vieta had found a means of solving equations allied
to evolution, and this was further developed by Harriot
and Oughtred, but without making the process less
tiresome.*   Harriot's name is connected with another
theorem which contains the law of formation of the
coefficients of an algebraic equation from its roots,
although the theorem was first stated in full by Descartes (1683) and proved general by Gauss.

Since there was lacking a sure method of determining the roots of equations of higher degree, the
attempt was made to include these roots within as
narrow limits as possible. De Beaune and Van
Schooten tried to do this, but the first usable methods
date from Maclaurin (*Algebra*, published posthumously in 1748) and Newton (1722) who fixed the real
roots of an algebraic equation between given limits.
In order to effect the general solution of an algebraic
equation, the effort was made either to represent the
given equation as the product of several equations of
lower degree, a method further developed by Hudde,
or to reduce, through extraction of the square root,
an equation of even degree to one whose degree is
half that of the given equation ; this method was used
by Newton, but he accomplished little in this direction.

Leibnitz had exerted himself as strenuously as
Newton to make advances in the theory of algebraic
equations.   In one of his letters he states that he has

* Montucla, *Histoire des Mathématiques*, 1799-1802.

been engaged for a long time in attempting to find
the irrational roots of an equation of any degree, by
eliminating the intermediate terms and reducing it to
the form $x^n = A$, and that he was persuaded that in
this manner the complete solution of the general equation of the nth degree could be effected. This method
of transformation of the general equation dates back
to Tschirnhausen and is found as "Nova methodus
etc." in the *Leipziger Acta eruditorum* of the year 1683.
In the equation

$$x^n + Ax^{n-1} + Bx^{n-2} + \ldots + Mx + N = 0$$

Tschirnhausen places

$$y = a + \beta x + \gamma x^2 + \ldots + \mu x^{n-1};$$

the elimination of x from these two equations gives
likewise an equation of the nth degree in y, in which
the undetermined coefficients a, β, γ, . . . can so be
taken as to give the equation in y certain special characteristics, for example, to make some of the terms
vanish. From the values of y, the values of x are determined. By this method the solution of equations
of the 3rd and 4th degrees is made to depend respectively upon those of the 2nd and 3rd degrees; but the
application of this method to the equation of the 5th
degree, leads to one of the 24th degree, upon whose
solution the complete solution of the equation of the
5th degree depends.

Afterwards, also, toward the end of the seventeenth
and the beginning of the eighteenth century, De Lagny,

Rolle, Laloubère, and Leseur made futile attempts to
advance with rigorous proofs beyond the equation of
the fourth degree.   Euler* took the problem in hand
in 1749.   He attempted first to resolve by means of
undetermined coefficients the equation of degree $2n$
into two equations each of degree n, but the results
obtained by him were not more satisfactory than those
of his predecessors, in that an equation of the eighth
degree by this treatment led to an equation of the 70th
degree. These investigations were not valueless, however, since through them Euler discovered the proof
of the theorem that every rational integral algebraic
function of even degree can be resolved into real factors of the second degree.

In a work of the date 1762 Euler attacked the solution of the equation of the nth degree directly. Judg-
ing from equations of the 2nd and 3rd degrees, he sur-
mised that a root of the general equation of the nth
degree might be composed of $(n-1)$ radicals of the
nth degree with subordinate square roots. He formed
expressions of this sort and sought through comparison of coefficients to accomplish his purpose.   This
method presented no difficulty up to the fourth degree, but in the case of the fifth degree Euler was
compelled to limit himself to particular cases.   For
example, he obtained from

$$x^5 - 40x^3 - 72x^2 + 50x + 98 = 0$$

the following value:

*Cantor, III., p. 582.

$$x = \sqrt[3]{-31 + 3\sqrt{-7}} + \sqrt[3]{-31 - 3\sqrt{-7}}$$
$$+ \sqrt[3]{-18 + 10\sqrt{-7}} + \sqrt[3]{-18 - 10\sqrt{-7}}.$$

Analogous to this attempt of Euler is that of Waring (1779). In order to solve the equation $f(x) = 0$ of degree n, he places

$$x = a\sqrt[n]{p} + b\sqrt[n]{p^2} + c\sqrt[n]{p^3} + \ldots + q\sqrt[n]{p^{n-1}}.$$

After clearing of radicals, he gets an equation of the nth degree, $F(x) = 0$, and by equating coefficients finds the necessary equations for determining a, b, c, $\ldots q$ and p, but is unable to complete the solution.

Bézout also proposed a method. He eliminated y from the equations $y^n - 1 = 0$, $ay^{n-1} + by^{n-2} + \ldots + x = 0$, and obtained an equation of the nth degree, $f(x) = 0$, and then equated coefficients. Bézout was no more able to solve the general equation of the 5th degree than Waring, but the problem gave him the impulse to perfect methods of elimination.

Tschirnhausen had begun, with his transformation, to study the roots of the general equation as functions of the coefficients. The same result can be reached by another method not different in principle, namely the formation of resolvents. In this way, Lagrange, Malfatti and Vandermonde independently reached results which were published in the year 1771. Lagrange's work, rich in matter, gave an analysis of all the then known methods of solving equations, and explained the difficulties which present themselves in passing beyond the fourth degree. Besides this he gave methods for determining the limits of the roots and the number of imaginary roots, as well as methods of approximation.

Thus all expedients for solving the general equation, made prior to the beginning of the nineteenth century yielded poor results, and especially with reference to Lagrange's work Montucla* says "all this is well calculated to cool the ardor of those who are inclined to tread this new way. Must one entirely despair of the solution of this problem?"

Since the general problem proved insoluble, attempts were made with special cases, and many elegant results were obtained in this way. De Moivre brought the solution of the equation

$$ny + \frac{n^2-1}{2\cdot 3}\, ny^3 + \frac{(n^2-1)\cdot(n^2-9)}{2\cdot 3\cdot 4\cdot 5}\, ny^5 + \ldots = a,$$

for odd integral values of n, into the form

$$y = \tfrac{1}{2}\sqrt[n]{a+\sqrt{a^2+1}} - \tfrac{1}{2}\sqrt[n]{-a-\sqrt{a^2+1}}.$$

Euler investigated symmetric equations and Bézout deduced the relation between the coefficients of an equation of the nth degree which must exist in order that the same may be transformed into $y^n + a = 0$.

Gauss made an especially significant step in advance in the solution of the cyclotomic equation $x^n - 1 = 0$, where n is a prime number. Equations of this sort are closely related to the division of the circum-

* *Histoire des Sciences Mathématiques*, 1799-1802.

ference into n equal parts. If y is the side of an inscribed n-gon in a circle of radius 1, and z the diagonal connecting the first and third vertices, then

$$y = 2 \sin \frac{\pi}{n}, \quad z = 2 \sin \frac{2\pi}{n}.$$

If however

$$x = \cos \frac{2\pi}{n} + i \sin \frac{2\pi}{n}, \quad \left(\cos \frac{2\pi}{n} + i \sin \frac{2\pi}{n} \right)^n = 1,$$

then the equation $x^n - 1 = 0$ is to be considered as the algebraic expression of the problem of the construction of the regular n-gon.

The following very general theorem was proved by Gauss.[*] "If n is a prime number, and if $n-1$ be resolved into prime factors $a, b, c, \ldots$ so that $n-1 = a^\alpha \, b^\beta \, c^\gamma \ldots$, then it is always possible to make the solution of $x^n - 1 = 0$ depend upon that of several equations of lower degree, namely upon α equations of degree a, β equations of degree b, etc." Thus for example, the solution of $x^{73} - 1 = 0$ (the division of the circumference into 73 equal parts) can be effected, since $n-1 = 72 = 3^2 . 2^3$, by solving three quadratic and two cubic equations. Similarly $x^{17} - 1 = 0$ leads to four equations of the second degree, since $n-1 = 16 = 2^4$; therefore the regular 17-gon can be constructed by elementary geometry, a fact which before the time of Gauss no one had anticipated.

Detailed constructions of the regular 17-gon by elementary geometry were first given by Pauker and

---

[*] Legendre, *Théorie des Nombres.*

Erchinger.[*]   A noteworthy construction of the same figure is due to von Staudt.

For the case that the prime number n has the form $2^m + 1$, the solution of the equation $x^n - 1 = 0$ depends upon the solution of m quadratic equations of which only $m - 1$ are necessary in the construction of the regular n-gon. It should be observed that for $m = 2^k$ (k a positive integer), the number $2^m + 1$ may be prime, but, as R. Baltzer[†] has pointed out, is not necessarily prime. If m is given successively the values

$$1, 2, 3, 4, 5, 6, 7, 8, 16, 2^{12}, 2^{23},$$

$n = 2^m + 1$ will take the respective values

$$3, 5, 9, 17, 33, 65, 129, 257, 65537, 2^{2^{12}} + 1, 2^{2^{23}} + 1,$$

of which only 3, 5, 17, 257, 65537 are prime. The remaining numbers are composite; in particular, the last two values of n have respectively the factors 114689 and 167772161. The circle therefore can be divided into 257 or 65537 equal parts by solving respectively 7 or 15 quadratic equations, which is possible by elementary geometric construction.

From the equalities

$$255 = 2^8 - 1 = (2^4 - 1)(2^4 + 1) = 15 \cdot 17, \qquad 256 = 2^8,$$
$$65535 = 2^{16} - 1 = (2^8 - 1)(2^8 + 1) = 255 \cdot 257, \quad 65536 = 2^{16},$$

it is easily seen that, by elementary geometry, that is, by use of only straight edge and compasses, the circle can be divided respectively into 255, 256, 257; 65535, 65536, 65537 equal parts. The process cannot be continued without a break, since $n = 2^{32} + 1$ is not prime.

The possibility of an elementary geometric construction of the regular 65535-gon is evident from the following:

$$65535 = 255 \cdot 257 = 15 \cdot 17 \cdot 257.$$

If the circumference of the circle is 1, then since

* Gauss, *Werke*, II., p. 187.

† Netto, *Substitutionentheorie*, 1882; English by Cole, 1892, p. 187.

$$\tfrac{1}{16} - \tfrac{1}{17} = \tfrac{1}{255}, \quad \tfrac{1}{255} - \tfrac{1}{257} = \tfrac{1}{65535}.$$

it follows that $\tfrac{1}{65535}$ of the circumference can be obtained by elementary geometric operations.

After Gauss had given in his earliest scientific work, his doctor's dissertation, the first of his proofs of the important theorem that every algebraic equation has a real or an imaginary root, he made in his great memoir of 1801 on the theory of numbers, the conjecture that it might be impossible to solve general equations of degree higher than the fourth by radicals. Ruffini and Abel gave a rigid proof of this fact, and it is due to these investigations that the fruitless efforts to reach the solution of the general equation by the algebraic method were brought to an end. In their stead the question formulated by Abel came to the front, "What are the equations of given degree which admit of algebraic solution?"

The cyclotomic equations of Gauss form such a group. But Abel made an important generalization by the theorem that an irreducible equation is always soluble by radicals when of two roots one can be rationally expressed in terms of the other, provided at the same time the degree of the equation is prime; if this is not the case, the solution depends upon the solution of equations of lower degree.

A further great group of algebraically soluble equations is therefore comprised in the Abelian equations. But the question as to the necessary and sufficient conditions for the algebraic solubility of an equation

was first answered by the youthful Galois, the crown
of whose investigations is the theorem, "If the degree
of an irreducible equation is a prime number, the
equation is soluble by radicals alone, provided the
roots of this equation can be expressed rationally in
terms of any two of them."

Abel's investigations fall between the years 1824
and 1829, those of Galois in the years 1830 and 1831.
Their fundamental significance for all further labors
in this direction is an undisputed fact; the question
concerning the general type of algebraically soluble
equations alone awaits an answer.

Galois, who also earned special honors in the field
of modular equations which enter into the theory of
elliptic functions, introduced the idea of a group of
substitutions.* The importance of this innovation,
and its development into a formal theory of substitutions, as Cauchy has first given it in the *Exercices
d'analyse*, etc., when he speaks of "systems of conjugate substitutions," became manifest through geometric considerations. The first example of this was
furnished by Hesse† in his investigation on the nine
points of inflexion of a curve of the third degree. The
equation of the ninth degree upon which they depend
belongs to the class of algebraically soluble equations.
In this equation there exists between any two of the
roots and a third determined by them an algebraic re-

---

* Netto, *Substitutionentheorie*, 1882. English by Cole, 1892.

† Noether, O. H., *Schlömilch's Zeitschrift*, Band 20.

lation expressing the geometric fact that the nine
points of inflexion lie by threes on twelve straight
lines. To the development of the substitution theory
in later times, Kronecker, Klein, Noether, Hermite,
Betti, Serret, Poincaré, Jordan, Capelli, and Sylow
especially have contributed.

Most of the algebraists of recent times have par-
ticipated in the attempt to solve the equation of the
fifth degree. Before the impossibility of the algebraic
solution was known, Jacobi at the age of 16 had made
an attempt in this direction; but an essential advance
is to be noted from the time when the solution of the
equation of the fifth degree was linked with the theory
of elliptic functions.* By the help of transformations
as given on the one hand by Tschirnhausen and on
the other by E. S. Bring (1786), the roots of the equation of the fifth degree can be made to depend upon
a single quantity only, and therefore the equation, as
shown by Hermite, can be put into the form $t^5 - t - A$
$= 0$. By Riemann's methods, the dependence of the
roots of the equation upon the parameter A is illustrated; on the other hand, it is possible by power-
series to calculate these five roots to any degree of approximation. In 1858, Hermite and Kronecker solved
the equation of the fifth degree by elliptic functions,
but without reference to the algebraic theory of this
equation, while Klein gave the simplest possible solu-

* Klein, F., *Vergleichende Betrachtungen über neuere geometrische Forschungen*, 1872.

tion by transcendental functions by using the theory of the icosahedron.

The solution of general equations of the nth degree for $n > 4$ by transcendental functions has therefore become possible, and the operations entering into the solution are the following: Solution of equations of lower degree; solution of linear differential equations with known singular points; determination of constants of integration, by calculating the moduli of periodicity of hyper-elliptic integrals for which the branch-points of the function to be integrated are known; finally the calculation of theta-functions of several variables for special values of the argument.

The methods leading to the complete solution of an algebraic equation are in many cases tedious; on this account the methods of approximation of real roots are very important, especially where they can be applied to transcendental equations. The most general method of approximation is due to Newton (communicated to Barrow in 1669), but was also reached by Halley and Raphson in another way.* For the solution of equations of the third and fourth degrees, John Bernoulli worked out a valuable method of approximation in his *Lectiones calculi integralis.* Further methods of approximation are due to Daniel Bernoulli, Taylor, Thomas Simpson, Lagrange, Legendre, Horner, and others.

By graphic and mechanical means also, the roots of an equation can be approximated. C. V. Boys† made use of a machine for this purpose, which consisted of a system of levers and fulcrums; Cunynghame‡ used a cubic parabola with a tangent scale

---

* Montucla.   † *Nature*, XXXIII., p. 166.

on a straight edge: C. Reuschle[*] used an hyperbola with an accompanying gelatine-sheet, so that the roots could be read as intersections of an hyperbola with a parabola. Similar methods,
suited especially to equations of the third and fourth degrees are
due to Bartl, R. Hoppe, and Oekinghaus[†]; Lalanne and Mehmke
also deserve mention in this connection.

For the solution of equations, there had been invented in the seventeenth century an algorism which
since then has gained a place in all branches of mathematics, the algorism of determinants.[‡] The first suggestion of computation with those regularly formed
aggregates, which are now called determinants (after
Cauchy), was given by Leibnitz in the year 1693.
He used the aggregate

$$a_{11}, a_{12}, \ldots \ldots a_{1n}$$
$$a_{21}, a_{22}, \ldots \ldots a_{2n}$$

in forming the resultant of n linear equations with
$n-1$ unknowns, and that of two algebraic equations
with one unknown. Cramer (1750) is considered as
a second inventor, because he began to develop a system of computation with determinants. Further theorems are due to Bézout (1764), Vandermonde (1771),
Laplace (1772), and Lagrange (1773). Gauss's *Disquisitiones arithmeticae* (1801) formed an essential ad-

[*] Böklen, O., *Math. Mittheilungen*, 1886, p. 102.

[†] *Fortschritte*, 1883; 1884.

[‡] Muir, T., *Theory of Determinants in the Historical Order of its Development*, Part I, 1890; Baltzer. R., *Theorie und Anwendungen der Determinanten*,
1881.

vance, and this gave Cauchy the impulse to many new investigations, especially the development of the general law (1812) of the multiplication of two determinants.

Jacobi by his "masterful skill in technique," also rendered conspicuous service in the theory of determinants, having developed a theory of expressions which he designated as "functional determinants." The analogy of these determinants with differential quotients led him to the general "principle of the last multiplier" which plays a part in nearly all problems of integration.* Hesse considered in an especially thorough manner symmetric determinants whose elements are linear functions of the co-ordinates of a geometric figure. He observed their behavior by linear transformation of the variables, and their relations to such determinants as are formed from them by a single bordering.† Later discussions are due to Cayley on skew determinants, and to Nachreiner and S. Günther on relations between determinants and continued fractions.

The appearance of the differential calculus forms one of the most magnificent discoveries of this period. The preparatory ideas for this discovery appear in manifest outline in Cavalieri,‡ who in a work *Methodus indivisibilium* (1635) considers a space-element as

* Dirichlet, "Gedächtnissrede auf Jacobi." *Crelle's Journal*, Band 52.
† Noether, O. H., *Schlömilch's Zeitschrift*, Band 20.
‡ Lüroth, *Rektoratsrede*, Freiburg, 1889; Cantor, II., p. 759.

the sum of an infinite number of simplest space-elements of the next lower dimension, e. g., a solid as the sum of an infinite number of planes. The danger of this conception was fully appreciated by the inventor of the method, but it was improved first by Pascal who considers a surface as composed of an infinite number of infinitely small rectangles, then by Fermat and Roberval; in all these methods, however, there appeared the drawback that the sum of the resulting series could seldom be determined. Kepler remarked that a function can vary only slightly in the vicinity of a greatest or least value. Fermat, led by this thought, made an attempt to determine the maximum or minimum of a function. Roberval investigated the problem of drawing a tangent to a curve, and solved it by generating the curved line by the composition of two motions, and applied the parallelogram of velocities to the construction of the tangents. Barrow, Newton's teacher, used this preparatory work with reference to Cartesian co-ordinate geometry. He chose the rectangle as the velocity-parallelogram, and at the same time introduced like Fermat infinitely small quantities as increments of the dependent and independent variables, with special symbols. He gave also the rule, that, without affecting the validity of the result of computation, higher powers of infinitely small quantities may be neglected in comparison with the first power. But Barrow was not able to handle fractions and radicals involving infinitely small quantities,

and was compelled to resort to transformations to re-
move them.   Like his predecessors, he was able to
determine in the simpler cases the value of the quotient of two, or the sum of an infinite number of infinitesimals.   The general solution of such questions
was reached by Leibnitz and Newton, the founders of
the differential calculus.

Leibnitz gave for the calculus of infinitesimals, the
notion of which had been already introduced, further
examples and also rules for more complicated cases.
By summation according to the old methods,* he deduced the simplest theorems of the integral calculus,
which he, by prefixing a long S as the sign of summation wrote,

$$\int x = \frac{x^2}{2}, \quad \int x^2 = \frac{x^3}{3}, \quad \int (x+y) = \int x + \int y.$$

From the fact that the sign of summation $\int$ raised
the dimension, he drew the conclusion that by differ-
ence-forming the dimension must be diminished so
that, therefore, as he wrote in a manuscript of Oct.
29, 1675, from $\int l = ya$, follows immediately $l = \frac{ya}{d}$.

Leibnitz tested the power of his new method by
geometric problems; he sought, for example, to determine the curve "for which the intercepts on the
axis to the feet of the normals vary as the ordinates."
In this he let the abscissas x increase in arithmetic
ratio and designated the constant difference of the

*Gerhardt, *Geschichte der Mathematik in Deutschland*, 1877; Cantor, III.,
p. 160.

abscissas first by $\frac{x}{d}$ and later by dx, without explaining in detail the meaning of this new symbol. In 1676 Leibnitz had developed his new calculus so far as to be able to solve geometric problems which could not be reduced by other methods. Not before 1686, however, did he publish anything about his method, its great importance being then immediately recognized by Jacob Bernoulli.

What Leibnitz failed to explain in the development of his methods, namely what is understood by his infinitely small quantities, was clearly expressed by Newton, and secured for him a theoretical superiority. Of a quotient of two infinitely small quantities Newton speaks as of a limiting value * which the ratio of the vanishing quantities approaches, the smaller they become. Similar considerations hold for the sum of an infinite number of such quantities. For the determination of limiting values, Newton devised an especial algorism, the calculus of fluxions, which is essentially identical with Leibnitz's differential calculus. Newton considered the change in the variable as a flowing; he sought to determine the velocity of the variation of the function when the variable changes with a given velocity. The velocities were called fluxions and were designated by $\dot{x}$, $\dot{y}$, $\dot{z}$ (instead of dx, dy, dz, as in Leibnitz's writings). The quantities themselves were called fluents, and the calculus of fluxions determines therefore the velocities of given

* Lüroth, *Rektoratsrede*, Freiburg, 1889.

motions, or seeks conversely to find the motions when
the law of their velocities is known. Newton's paper
on this subject was finished in 1671 under the name
of *Methodus fluxionum,* but was first published in 1736,
after his death. Newton is thought by some to have
borrowed the idea of fluxions from a work of Napier.*

According to Gauss, Newton deserved much more
credit than Leibnitz, although he attributes to the
latter great talent, which, however, was too much dis-
sipated. It appears that this judgment, looked at
from both sides, is hardly warranted. Leibnitz failed
to give satisfactory explanation of that which led
Newton to one of his most important innovations, the
idea of limits. On the other hand, Newton is not
always entirely clear in the purely analytic proo .
Leibnitz, too, deserves very high praise for the introduction of the appropriate symbols $\int$ and dx, as well
as for stating the rules of operating with them. To-
day the opinion might safely be expressed that the
differential and integral calculus was independently
discovered by Newton and by Leibnitz; that Newton
is without doubt the first inventor; that Leibnitz, on
the other hand, stimulated by the results communicated to him by Newton, but without the knowledge
of Newton's methods, invented independently the
calculus; and that finally to Leibnitz belongs the
priority of publication."†

* Cohen, *Das Prinzip der Infinitesimalmethode und seine Geschichte,* 1883;
Cantor, III., p. 163.
† Lüroth. A very good summary of the discussion is also given in Ball's

The systematic development of the new calculus made necessary a clearer understanding of the idea of the infinite. Investigations on the infinitely great are of course of only passing interest for the explanation of natural phenomena,* but it is entirely different with the question of the infinitely small. The infinitesimal† appears in the writings of Kepler as well as in those of Cavalieri and Wallis under varying forms, essentially as "infinitely small null-value," that is, as a quantity which is smaller than any given quantity, and which forms the limit of a given finite quantity. Euler's *indivisibilia* lead systematically in the same direction. Fermat, Roberval, Pascal, and especially Leibnitz and Newton operated with the "unlimitedly small," yet in such a way that frequently an abbreviated method of expression concealed or at least obscured the true sense of the development. In the writings of John Bernoulli, De l'Hospital, and Poisson, the infinitesimal appears as a quantity different from zero, but which must become less than an assignable value, i. e., as a " pseudo-infinitesimal" quantity. By the formation of derivatives, which in the main are identical with Newton's fluxions, Lagrange‡ attempted entirely to avoid the infinitesimal, but his attempts only served the purpose of bringing into

Short History of Mathematics, London, 1888. The best summary is that given in Cantor, Vol. III.

* Riemann, *Werke*, p. 267.

† R. Hoppe, *Differentialrechnung*, 1865.

‡ Lüroth, *Rektoratsrede*, Freiburg, 1889.

prominence the urgent need for a deeper foundation
for the theory of the infinitesimal for which Tacquet
and Pascal in the seventeenth century, and Maclaurin
and Carnot in the eighteenth had made preparation.
We are indebted to Cauchy for this contribution.   In
his investigations there is clearly established the meaning of propositions which contain the expression "infinitesimal," and a safe foundation for the differential
calculus is thereby laid.

The integral calculus was first further extended
by Cotes, who showed how to integrate rational algebraic functions.   Legendre applied himself to the integration of series, Gauss to the approximate determination of integrals, and Jacobi to the reduction and
evaluation of multiple integrals.   Dirichlet is especially to be credited with generalizations on definite
integrals, his lectures showing his great fondness for
this theory.*   He it was who welded the scattered
results of his predecessors into a connected whole,
and enriched them by a new and original method of
integration.   The introduction of a discontinuous factor allowed him to replace the given limits of integration by different ones, often by infinite limits, without
changing the value of the integral.   In the more re
cent investigations the integral has become the means
of defining functions or of generating others.

In the realm of differential equations† the works

---

* Kummer, "Gedächtnissrede auf Lejeune-Dirichlet." *Berliner Abh.*, 1860
† Cantor, III., p. 429; Schlesinger, L., *Handbuch der Theorie der linearen*

worthy of mention date back to Jacob and John Bernoulli and to Riccati. Riccati's merit consists mainly in having introduced Newton's philosophy into Italy. He also integrated for special cases the differential equation named in his honor—an equation completely solved by Daniel Bernoulli—and discussed the question of the possibility of lowering the order of a given differential equation. The theory first received a detailed scientific treatment at the hands of Lagrange, especially as far as concerns partial differential equations, of which D'Alembert and Euler had handled the equation $\frac{d^2u}{dt^2} = \frac{d^2u}{dx^2}$. Laplace also wrote on this differential equation and on the reduction of the solution of linear differential equations to definite integrals.

On German soil, J. F. Pfaff, the friend of Gauss and next to him the most eminent mathematician of that time, presented certain elegant investigations (1814, 1815) on differential equations,[*] which led Jacobi to introduce the name "Pfaffian problem." Pfaff found in an original way the general integration of partial differential equations of the first degree for any number of variable quantities. Beginning with the theory of ordinary differential equations of the first degree with n variables, for which integrations

Differentialgleichungen, Bd. I., 1895,—an excellent historical review; Mansion, P., *Theorie der partiellen Differentialgleichungen erster Ordnung*, deutsch von Maser, Leipzig, 1892, also excellent on history.

[*]A. Brill, "Das mathematisch-physikalische Seminar in Tübingen." Aus der *Festschrift der Universität zum Königs-Jubiläum*, 1889.

were given by Monge (1809) in special simple cases,
Pfaff gave their general integration and considered
the integration of partial differential equations as a
particular case of the general integration. In this the
general integration of differential equations of every
degree between two variables is assumed as known.*
Jacobi (1827, 1836) also advanced the theory of differential equations of the first order. The treatment
was so to determine unknown functions that an integral which contains these functions and the differential
coefficient in a prescribed way reaches a maximum or
minimum. The condition therefor is the vanishing of.
the first variation of the integral, which again finds its
expression in differential equations, from which the
unknown functions are determined. In order to be
able to distinguish whether a real maximum or minimum appears, it is necessary to bring the second variation into a form suitable for investigating its sign.
This leads to new differential equations which Lagrange was not able to solve, but of which Jacobi was
able to show that their integration can be deduced
from the integration of differential equations belonging to the first variation. Jacobi also investigated
the special case of a simple integral with one unknown
function, his statements being completely proved by
Hesse. Clebsch undertook the general investigation
of the second variation, and he was successful in
showing for the case of multiple integrals that new

*Gauss, *Werke*, III., p. 232.

integrals are not necessary for the reduction of the second variation. Clebsch (1861, 1862), following the suggestions of Jacobi, also reached the solution of the Pfaffian problem by making it depend upon a system of simultaneous linear partial differential equations, the statement of which is possible without integration. Of other investigations, one of the most important is the theory of the equation

$$\frac{d^2v}{dx^2} + \frac{d^2v}{dy^2} + \frac{d^2v}{dz^2} = 0,$$

which Dirichlet encountered in his work on the potential, but which had been known since Laplace (1789). Recent investigations on differential equations, especially on the linear by Fuchs, Klein, and Poincaré, stand in close connection with the theories of functions and groups, as well as with those of equations and series.

"Within a half century the theory of ordinary differential equations has come to be one of the most important branches of analysis, the theory of partial differential equations remaining as one still to be perfected. The difficulties of the general problem of integration are so manifest that all classes of investigators have confined themselves to the properties of the integrals in the neighborhood of certain given points. The new departure took its greatest inspiration from two memoirs by Fuchs (1866, 1868), a work elaborated by Thomé and Frobenius . . .

"Since 1870 Lie's labors have put the entire theory of differential equations on a more satisfactory foundation. He has shown that the integration theories of the older mathematicians, which had been looked upon as isolated, can by the introduction of the concept of continuous groups of transformations be referred to a common source, and that ordinary differential equations which admit the same infinitesimal transformations present like difficulties of integration   He has also emphasized the subject of transformations of contact (*Berührungs-Transformationen*) which underlies so much of the recent theory. . . . Recent writers have shown the same tendency noticeable in the works of Monge and Cauchy, the tendency to separate into two schools, the one inclining to use the geometric diagram and represented by Schwarz, Klein, and Goursat, the other adhering to pure analysis, of which Weierstrass, Fuchs, and Frobenins are types."[*]

A short time after the discovery of the differential and integral calculus, namely in the year 1696, John Bernoulli proposed this problem to the mathematicians of his time: To find the curve described by a body falling from a given point A to another given point B in the shortest time.[†] The problem came from a case in optics, and requires a function to be found whose integral is a minimum. Huygens had developed the wave-theory of light, and John Bernoulli had found under definite assumptions the differential equation of the path of the ray of light. Of such motion he sought another example, and came upon the cycloid as the brachistochrone, that is, upon the above statement of the problem, for which up to Easter 1697, solutions from the Marquis de l'Hospital, from Tschirnhausen, Newton, Jacob Bernoulli and Leibnitz were received. Only the two latter treated the

[*] Smith, D. E., "History of Modern Mathematics," in Merriman and Woodward's *Higher Mathematics*, New York, 1896, with authorities cited.

[†] Reiff, R., "Die Anfänge der Variationsrechnung," *Math. Mittheilungen von Böklen*, 1887. Cantor, III., p. 225. Woodhouse, *A Treatise on Isoperimetrical Problems* (Cambridge, 1810). The last named work is rare.

problem as one of maxima and minima. Jacob Bernoulli's method remained the common one for the treatment of similar cases up to the time of Lagrange, and he is therefore to be regarded as one of the founders of the calculus of variations. At that time[*] all problems which demanded the statement of a maximum or minimum property of functions were called isoperimetric problems. To the oldest problems of this kind belong especially those in which one curve with a maximum or minimum property was to be found from a class of curves of equal perimeters. That the circle, of all isoperimetric figures, gives the maximum area, is said to have been known to Pythagoras. In the writings of Pappus a series of propositions on figures of equal perimeters are found. Also in the fourteenth century the Italian mathematicians had worked on problems of this kind. But "the calculus of variations may be said to begin with . . . John Bernoulli (1696). It immediately occupied the attention of Jacob Bernoulli and the Marquis de l'Hospital, but Euler first elaborated the subject."[†] He[‡] investigated the isoperimetric problem first in the analytic-geometric manner of Jacob Bernoulli, but after he had worked on the subject eight years, he came in 1744 upon a new and general solution by a purely analytic method (in his celebrated work: *Methodus inveniendi*

[*] Anton, *Geschichte des isoperimetrischen Problems*, 1888.
[†] Smith, D. E., *History of Modern Mathematics*, p. 533.
[‡] Cantor, III., pp. 243, 819, 830.

lineas curvas, etc.); this solution shows how those ordinates of the function which are to assume a greatest
or least value can be derived from the variation of the
curve-ordinate. Lagrange (*Essai d'une nouvelle méthode*, etc., 1760 and 1761) made the last essential step
from the pointwise variation of Euler and his predecessors to the simultaneous variation of all ordinates
of the required curve by the assumption of variable
limits of the integral. His methods, which contained
the new feature of introducing δ for the change of the
function, were later taken up in Euler's *Integral Calculus*. Since then the calculus of variations has been
of valuable service in the solution of problems in theory of curvature.

The beginnings of a real theory of functions*, especially that of the elliptic and Abelian functions lead
back to Fagnano, Maclaurin, D'Alembert, and Landen.
Integrals of irrational algebraic functions were treated,
especially those involving square roots of polynomials
of the third and fourth degrees; but none of these
works hinted at containing the beginnings of a science
dominating the whole subject of algebra. The matter
assumed more definite form under the hands of Euler,
Lagrange, and Legendre. For a long time the only
transcendental functions known were the circular func-

---

*Brill, A., and Noether, M., "Die Entwickelung der Theorie der alge-
braischen Functionen in älterer und neuerer Zeit, Bericht erstattet der Deutschen Mathematiker-Vereinigung. *Jahresbericht*, Bd. II., pp. 107-566, Berlin,
1894 ; Königsberger, L., *Zur Geschichte der Theorie der elliptischen Transcen-
denten in den Jahren 1826-1829*, Leipzig, 1879.

tions (sin x, cos x, . . .), the common logarithm, and, especially for analytic purposes, the hyperbolic logarithm with base e, and (contained in this) the exponential function e^x. But with the opening of the nineteenth century mathematicians began on the one hand thoroughly to study special transcendental functions, as was done by Legendre, Jacobi, and Abel, and on the other hand to develop the general theory of functions of a complex variable, in which field Gauss, Cauchy, Dirichlet, Riemann, Liouville, Fuchs, and Weierstrass obtained valuable results.

The first signs of an interest in elliptic functions[*] are connected with the determination of the arc of the lemniscate, as this was carried out in the middle of the eighteenth century. In this Fagnano made the discovery that between the limits of two integrals expressing the arc of the curve, one of which has twice the value of the other, there exists an algebraic relation of simple nature. By this means, the arc of the lemniscate, though a transcendent of higher order, can be doubled or bisected by geometric construction like an arc of a circle.[†] Euler gave the explanation of this remarkable phenomenon. He produced a more general integral than Fagnano (the so-called elliptic integral of the first class) and showed that two such integrals can be combined into a third of the same kind, so that between the limits of these

---

[*] Enneper, A., *Elliptische Functionen, Theorie und Geschichte*, Halle, 1890.

[†] Dirichlet, "Gedächtnissrede auf Jacobi." *Crelle's Journal*, Bd. 52.

integrals there exists a simple algebraic relation, just
as the sine of the sum of two arcs can be composed of
the same functions of the separate arcs (addition-theorem).   The elliptic integral, however, depends not
merely upon the limits but upon another quantity belonging to the function, the modulus.   While Euler
placed only integrals with the same modulus in relation, Landen and Lagrange considered those with
different moduli, and showed that it is possible by
simple algebraic substitution to change one elliptic
integral into another of the same class.   The establishment of the addition-theorem will always remain
at least as important a service of Euler as his transformation of the theory of circular functions by the
introduction of imaginary exponential quantities.

The origin* of the real theory of elliptic functions
and the theta-functions falls between 1811 and 1829.
To Legendre are due two systematic works, the *Exercices de calcul intégral* (1811–1816) and the *Théorie des
fonctions elliptiques* (1825–1828), neither of which was
known to Jacobi and Abel.   Jacobi published in 1829
the *Fundamenta nova theoriae functionum ellipticarum*,
certain of the results of which had been simultaneously discovered by Abel.   Legendre had recognised
that a new branch of analysis was involved in those
investigations, and he devoted decades of earnest
work to its development. Beginning with the integral
which depends upon a square root of an expression of

* Cayley, *Address to the British Association*, etc., 1883.

the fourth degree in x, Legendre noticed that such integrals can be reduced to canonical forms. $\Delta\psi = \sqrt{1-k^2\sin^2\psi}$ was substituted for the radical, and three essentially different classes of elliptic integrals were distinguished and represented by $F(\psi)$, $E(\psi)$, $\Pi(\psi)$. These classes depend upon the amplitude ψ and the modulus k, the last class also upon a parameter n.

In spite of the elegant investigations of Legendre on elliptic integrals, their theory still presented sev eral enigmatic phenomena. It was noticed that the degree of the equation conditioning the division of the elliptic integral is not equal to the number of the parts, as in the division of the circle, but to its square. The solution of this and similar problems was reserved for Jacobi and Abel. Of the many productive ideas of these two eminent mathematicians there are especially two which belong to both and have greatly advanced the theory.

In the first place, Abel and Jacobi independently of each other observed that it is not expedient to investigate the elliptic integral of the first class as a function of its limits, but that the method of consideration must be reversed, and the limit introduced as a function of two quantities dependent upon it. Expressed in other words, Abel and Jacobi introduced the direct functions instead of the inverse. Abel called them ϕ, f, F, and Jacobi named them *sin am*, *cos am*, Δ *am*, or, as they are written by Gudermann, *sn*, *cn*, *dn*.

A second ingenious idea, which belongs to Jacobi
as well as to Abel, is the introduction of the imaginary into this theory.   As Jacobi himself affirmed, it
was just this innovation which rendered possible the
solution of the enigma of the earlier theory. It turned
out that the new functions partake of the nature of
the trigonometric and exponential functions.   While
the former are periodic only for real values of the argument, and the latter only for imaginary values, the
elliptic functions have two periods.   It can safely be
said that Gauss as early as the beginning of the nineteenth century had recognised the principle of the
double period, a fact which was first made plain in
the writings of Abel.

Beginning with these two fundamental ideas, Jacobi and Abel, each in his own way, made further
important contributions to the theory of elliptic functions.   Legendre had given a transformation of one
elliptic integral into another of the same form, but a
second transformation discovered by him was unknown to Jacobi, as the latter after serious difficulties
reached the important result that a multiplication in
the theory of such functions can be composed of two
transformations.   Abel applied himself to problems
concerning the division and multiplication of elliptic
integrals. A thorough study of double periodicity led
him to the discovery that the general division of the
elliptic integral with a given limit is always algebraically possible as soon as the division of the complete

integrals is assumed as accomplished. The solution of the problem was applied by Abel to the lemniscate, and in this connection it was proved that the division of the whole lemniscate is altogether analogous to that of the circle, and can be performed algebraically in the same case. Another important discovery of Abel's occurred in his allowing, for elliptic functions of multiple argument, the multiplier to become infinite in formulas deduced from functions with a single argument. From this resulted the remarkable expressions which represent elliptic functions by infinite series or quotients of infinite products.

Jacobi had assumed in his investigations on transformations that the original variable is rationally expressible in terms of the new. Abel, however, entered this field with the more general assumption that between these two quantities an algebraic equation exists, and the result of his labor was that this more general problem can be solved by the help of the special problem completely treated by Jacobi.

Jacobi carried still further many of the investigations of Abel. Abel had given the theory of the general division, but the actual application demanded the formation of certain symmetric functions of the roots which could be obtained only in special cases. Jacobi gave the solution of the problem so that the required functions of the roots could be obtained at once and in a manner simpler than Abel's. When Jacobi had reached this goal, he stood alone on the broad expanse of the new science, for Abel a short time before had found an early grave at the age of 27.

The later efforts of Jacobi culminate in the introduction of the theta-function. Abel had already represented elliptic functions as quotients of infinite products. Jacobi could represent these products as special cases of a single transcendent, a fact which the French mathematicians had come upon in physical researches but had neglected to investigate. Jacobi examined their analytic nature, brought them into connection with the integrals of the second and third class, and noticed especially that integrals of the third class, though dependent upon three elements, can be represented by means of the new transcendent involving only two elements. The execution of this process gave to the whole theory a high degree of comprehensiveness and clearness, allowing the elliptic functions *sn, cn, dn* to be represented with the new Jacobian transcendents Θ_1, Θ_2, Θ_3, Θ_4 as fractions having a common denominator.

What Abel accomplished in the theory of elliptic functions is conspicuous, although it was not his greatest achievement. There is high authority for saying that the achievements of Abel were as great in the algebraic field as in that of elliptic functions. But his most brilliant results were obtained in the theory of the Abelian functions named in his honor, their first development falling in the years 1826–1829. "Abel's Theorem" has been presented by its discoverer in different forms. The paper, *Mémoire sur une propriété générale d'une classe très-étendue de fonctions transcendentes*, which after the death of the author received the prize from the French academy, contained the most general expression. In form it is a theorem of the integral calculus, the integrals depending upon an irrational function y, which is connected with x by an algebraic equation $F(x, y) = 0$. Abel's fundamental theorem states that a sum of such integrals can be expressed by a definite number p of similar integrals where p depends only upon the properties of the equation $F(x, y) = 0$. (This p is the deficiency of the curve $F(x, y) = 0$; the notion of deficiency, however, dates first from the year 1857.) For the case that

$$y = \sqrt{Ax^4 + Bx^3 + Cx^2 + Dx + E},$$

Abel's theorem leads to Legendre's proposition on the sum of two elliptic integrals. Here $p = 1$. If

$$y = \sqrt{Ax^6 + Bx^5 + \ldots + P},$$

where A can also be 0, then p is 2, and so on. For $p = 3$, or > 3, the hyperelliptic integrals are only special cases of the Abelian integrals of like class.

After Abel's death (1829) Jacobi carried the theory further in his *Considerationes generales de transcendentibus Abelianis* (1832), and showed for hyperelliptic integrals of a given class that the direct functions to which Abel's proposition applies are not functions of a single variable, as the elliptic functions *sn*, *cn*, *dn*, but are functions of p variables. Separate papers of

essential significance for the case $p=2$, are due to
Rosenhain (1846, published 1851) and Goepel (1847).

Two articles of Riemann, founded upon the writings of Gauss and Cauchy, have become significant
in the development of the complete theory of functions. Cauchy had by rigorous methods and by the
introduction of the imaginary variable "laid the foundation for an essential improvement and transformation of the whole of analysis."* Riemann built upon
this foundation and wrote the *Grundlage für eine allgemeine Theorie der Funktionen einer veränderlichen
komplexen Grösse* in the year 1851, and the *Theorie der
Abel'schen Funktionen* which appeared six years later.
For the treatment of the Abelian functions, Riemann
used theta-functions with several arguments, the theory of which is based upon the general principle of
the theory of functions of a complex variable. He
begins with integrals of algebraic functions of the
most general form and considers their inverse functions, that is, the Abelian functions of p variables.
Then a theta function of p variables is defined as the
sum of a p-tuply infinite exponential series whose
general term depends, in addition to p variables, upon
certain $\dfrac{p(p-1)}{2}$ constants which must be reducible
to $3p-3$ moduli, but the theory has not yet been completed.

Starting from the works of Gauss and Abel as well

* Kummer, "Gedächtnissrede auf Lejeune-Dirichlet," *Berliner Abhandlungen*, 1860.

as the developments of Cauchy on integrations in the
imaginary plane, a strong movement appears in which
occur the names of Weierstrass, G. Cantor, Heine,
Dedekind, P. Du Bois-Reymond, Dini, Scheeffer,
Pringsheim, Hölder, Pincherle, and others. This
tendency aims at freeing from criticism the foundations of arithmetic, especially by a new treatment of
irrationals based upon the theory of functions with its
considerations of continuity and discontinuity. It
likewise considers the bases of the theory of series by
investigations on convergence and divergence, and
gives to the differential calculus greater preciseness
through the introduction of mean-value theorems.

After Riemann valuable contributions to the theory
of the theta-functions were made by Weierstrass,
Weber, Nöther, H. Stahl, Schottky, and Frobenius.
Since Riemann a theory of algebraic functions and
point-groups has been detached from the theory of
Abelian functions, a theory which was founded through
the writings of Brill, Nöther, and Lindemann upon
the remainder-theorem and the Riemann-Roch theorem, while recently Weber and Dedekind have allied
themselves with the theory of ideal numbers, set forth
in the first appendix to Dirichlet. The extremely
rich development of the general theory of functions
in recent years has borne fruit in different branches of
mathematical science, and undoubtedly is to be recognised as having furnished a solid foundation for the
work of the future.

IV. GEOMETRY.

A. GENERAL SURVEY.

THE oldest traces of geometry are found among the Egyptians and Babylonians. In this first period geometry was made to serve practical purposes almost exclusively. From the Egyptian and Babylonian priesthood and learned classes geometry was transplanted to Grecian soil. Here begins the second period, a classic era of philosophic conception of geometric notions as the embodiment of a general science of mathematics, connected with the names of Pythagoras, Eratosthenes, Euclid, Apollonius, and Archimedes. The works of the last two indeed, touch upon lines not clearly defined until modern times. Apollonius in his *Conic Sections* gives the first real example of a geometry of position, while Archimedes for the most part concerns himself with the geometry of measurement.

The golden age of Greek geometry was brief and yet it was not wholly extinct until the memory of the great men of Alexandria was lost in the insignificance of their successors. Then followed more than a thousand years of a cheerless epoch which at best was restricted to borrowing from the Greeks such geometric knowledge as could be understood. History might pass over these many centuries in silence were it not compelled to give attention to these obscure and unproductive periods in their relation to the past and future. In this third period come first the Romans, Hindus, and Chinese, turning the Greek geometry to use after their own fashion; then the Arabs as skilled intermediaries between the ancient classic and a modern era.

The fourth period comprises the early development of geometry among the nations of the West. By the labors of Arab authors the treasures of a time long past were brought within the walls of monasteries and into the hands of teachers in newly established schools and universities, without as yet forming a subject for general instruction. The most prominent intellects of this period are Vieta and Kepler. In their methods they suggest the fifth period which begins with Descartes. The powerful methods of analysis are now introduced into geometry. Analytic geometry comes into being. The application of its seductive methods received the almost exclusive attention of the mathematicians of the seventeenth and eighteenth centuries. Then in the so-called modern or projective geometry and the geometry of curved surfaces there arose theories which, like analytic geometry, far transcended the geometry of the ancients, especially in

the way of leading to the almost unlimited generalization of truths already known.

B. FIRST PERIOD.

EGYPTIANS AND BABYLONIANS.

In the same book of Ahmes which has disclosed to
us the elementary arithmetic of the Egyptians are
also found sections on geometry, the determination
of areas of simple surfaces, with figures appended.
These figures are either rectilinear or circular. Among
them are found isosceles triangles, rectangles, isosceles trapezoids and circles.* The area of the rectangle is correctly determined; as the measure of the
area of the isosceles triangle with base a and side b,
however, $\frac{1}{2}ab$ is found, and for the area of the isosceles
trapezoid with parallel sides a' and a'' and oblique side
b, the expression $\frac{1}{2}(a' + a'')b$ is given. These approximate formulae are used throughout and are evidently
considered perfectly correct. The area of the circle
follows, with the exceptionally accurate value $\pi =
\left(\frac{16}{9}\right)^2 = 3.1605$.

Among the problems of geometric construction
one stands forth preeminent by reason of its practical
importance, viz., to lay off a right angle. The solution of this problem, so vital in the construction of
temples and palaces, belonged to the profession of

*Cantor, I., p. 52.

rope-stretchers or harpedonaptae. They used a rope
divided by knots into three segments (perhaps corresponding to the numbers 3, 4, 5) forming a Pythago-
ean triangle.*

Among the Babylonians the construction of figures
of religious significance led up to a formal geometry of
divination which recognized triangles, quadrilaterals,
right angles, circles with the inscribed regular hexagon and the division of the circumference into three
hundred and sixty degrees as well as a value $\pi = 3$.

Stereometric problems, such as finding the contents of granaries, are found in Ahmes; but not much
is to be learned from his statements since no account
is given of the shape of the storehouses.

As for projective representations, the Egyptian
wall-sculptures show no evidence of any knowledge
of perspective. For example a square pond is pictured in the ground-plan but the trees and the water-
drawers standing on the bank are added to the picture
in the elevation, as it were from the outside.†

C. SECOND PERIOD.

THE GREEKS.

In a survey of Greek geometry it will here and
there appear as if investigations connected in a very

---

* Cantor, I., p. 62.

† Wiener, *Lehrbuch der darstellenden Geometrie*, 1884. Hereafter referred
to as Wiener.

simple manner with well-known theorems were not
known to the Greeks.   At least it seems as if they
could not have been established satisfactorily, since
they are thrown in among other matters evidently
without connection.   Doubtless the principal reason
for this is that a number of the important writings of
the ancient mathematicians are lost.   Another no less
weighty reason might be that much was handed down
simply by oral tradition, and the latter, by reason of
the stiff and repulsive way in which most of the Greek
demonstrations were worked out, did not always render the truths set forth indisputable.

In Thales are found traces of Egyptian geometry,
but one must not expect to discover there all that was
known to the Egyptians.   Thales mentions the theorems regarding vertical angles, the angles at the base
of an isosceles triangle, the determination of a triangle
from a side and two adjacent angles, and the angle inscribed in a semi-circle.   He knew how to determine
the height of an object by comparing its shadow with
the shadow of a staff placed at the extremity of the
shadow of the object, so that here may be found the
beginnings of the theory of similarity.   In Thales the
proofs of the theorems are either not given at all or
are given without the rigor demanded in later times.

In this direction an important advance was made
by Pythagoras and his school.   To him without question is to be ascribed the theorem known to the Egyptian "rope-stretchers" concerning the right-angled

triangle, which they knew in the case of the triangle with sides 3, 4, 5, without giving a rigorous
proof. Euclid's is the earliest of the extant proofs of
this theorem. Of other matters, what is to be ascribed
to Pythagoras himself, and what to his pupils, it is
difficult to decide. The Pythagoreans proved that the
sum of the angles of a plane triangle is two right angles. They knew the golden section, and also the
regular polygons so far as they make up the boundaries of the five regular bodies. Also regular star-
polygons were known, at least the star-pentagon. In
the Pythagorean theorems of area the gnomon played
an important part. This word originally signified the
vertical staff which by its shadow indicated the hours,
and later the right angle mechanically represented.
Among the Pythagoreans the gnomon is the figure
left after a square has been taken from the corner of
another square. Later, in Euclid, the gnomon is a
parallelogram after similar treatment (see page 66).
The Pythagoreans called the perpendicular to a straight
line "a line directed according to the gnomon."*

But geometric knowledge extended beyond the
school of Pythagoras. Anaxagoras is said to have been
the first to try to determine a square of area equal
to that of a given circle. It is to be noticed that like
most of his successors he believed in the possibility
of solving this problem. Œnopides showed how to
draw a perpendicular from a point to a line and how

* Cantor, I., p. 150.

to lay off a given angle at a given point of a given line.   Hippias of Elis likewise sought the quadrature of the circle, and later he attempted the trisection of an angle, for which he constructed the quadratrix.

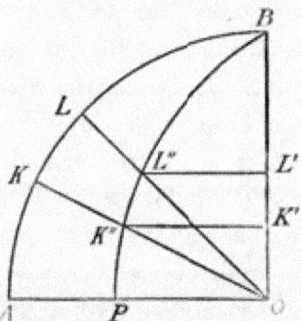

This curve is described as follows: Upon a quadrant of a circumference cut off by two perpendicular radii, OA and OB, lie the points $A, \ldots K, L, \ldots B$. The radius $r = OA$ revolves with uniform velocity about O from the position OA to the position OB. At the same time a straight line g always parallel to OA moves with uniform velocity from the position OA to that of a tangent to the circle at B. If K' is the intersection of g with OB at the time when the moving radius falls upon OK then the parallel to OA through K' meets the radius OK in a point K'' belonging to the quadratrix. If P is the intersection of OA with the quadratrix, it follows in part directly and in part from simple considerations, that

$$\frac{\text{arc } AK}{\text{arc } AL} = \frac{OK'}{OL'},$$

a relation which solves any problem of angle sections.   Furthermore,

$$OP = \frac{2r}{\pi}, \text{ or } \frac{OP}{OA} = \frac{OA}{\text{arc } AB},$$

whence it is obvious that the quadrature of the circle depends upon the ratio in which the radius OA is divided by the point P of the quadratrix. If this ratio could be constructed by elementary geometry, the quadrature of the circle would be effected.* It appears that the quadratrix was first invented for the trisection of an angle and that its relation to the quadrature of the circle was discovered later,† as is shown by Dinostratus.

The problem of the quadrature of the circle is also found in Hippocrates. He endeavored to accomplish his purpose by the consideration of crescent-shaped figures bounded by arcs of circles. It is of especial importance to note that Hippocrates wrote an elementary book of mathematics (the first of the kind) in which he represented a point by a single capital letter and a segment by two, although we are unable to determine who was the first to introduce this symbolism.

Geometry was strengthened on the philosophic side by Plato, who felt the need of establishing definitions and axioms and simplifying the work of the investigator by the introduction of the analytic method.

A systematic representation of the results of all the earlier investigations in the domain of elementary geometry, enriched by the fruits of his own abundant labor, is given by Euclid in the thirteen books of his *Elements* which deal not only with plane figures but also with figures in space and algebraic investiga-

* The equation of the quadratrix in polar co-ordinates is $r = \frac{2\phi}{\pi} \cdot \frac{a}{\sin\phi}$, where $a = OA$. Putting $\phi = 0$, $r = r_0$, we have $\pi = \frac{2a}{r_0}$.

† Montucla.

tions. "Whatever has been said in praise of mathematics, of the strength, perspicuity and rigor of its presentation, all is especially true of this work of the great Alexandrian. Definitions, axioms, and conclusions are joined together link by link as into a chain, firm and inflexible, of binding force but also cold and hard, repellent to a productive mind and affording no room for independent activity. A ripened understanding is needed to appreciate the classic beauties of this greatest monument of Greek ingenuity. It is not the arena for the youth eager for enterprise; to captivate him a field of action is better suited where he may hope to discover something new, unexpected."[*]

The first book of the *Elements* deals with the theory of triangles and quadrilaterals, the second book with the application of the Pythagorean theorem to a large number of constructions, really of arithmetic nature. The third book introduces circles, the fourth book inscribed and circumscribed polygons. Proportions explained by the aid of line-segments occupy the fifth book, and in the sixth book find their application to the proof of theorems involving the similarity of figures. The seventh, eighth, ninth and tenth books have especially to do with the theory of numbers. These books contain respectively the measurement and division of numbers, the algorism for determining the least common multiple and the greatest common divisor, prime numbers, geometric series,

[*] A. Brill, *Antrittsrede in Tübingen*, 1884.

and incommensurable (irrational) numbers. Then follows stereometry: in the eleventh book the straight line, the plane, the prism; in the twelfth, the discussion of the prism, pyramid, cone, cylinder, sphere; and in the thirteenth, regular polygons with the regular solids formed from them, the number of which Euclid gives definitely as five. Without detracting in the least from the glory due to Euclid for the composition of this imperishable work, it may be assumed that individual portions grew out of the well grounded preparatory work of others. This is almost certainly true of the fifth book, of which Eudoxus seems to have been the real author.

Not by reason of a great compilation like Euclid, but through a series of valuable single treatises, Archimedes is justly entitled to have a more detailed description of his geometric productions. In his investigations of the sphere and cylinder he assumes that the straight line is the shortest distance between two points. From the Arabic we have a small geometric work of Archimedes consisting of fifteen so-called lemmas, some of which have value in connection with the comparison of figures bounded by straight lines and arcs of circles, the trisection of the angle, and the determination of cross-ratios. Of especial importance is his mensuration of the circle, in which he shows π to lie between $3\frac{1}{7}$ and $3\frac{10}{71}$. This as well as many other results Archimedes obtains by the method of exhaustions which among the ancients usually took

the place of the modern integration.* The quantity
sought, the area bounded by a curve, for example,
may be considered as the limit of the areas of the inscribed and circumscribed polygons the number of
whose sides is continually increased by the bisection
of the arcs, and it is shown that the difference between
two associated polygons, by an indefinite continuance
of this process, must become less than an arbitrarily
small given magnitude. This difference was thus, as
it were, exhausted, and the result obtained by exhaus-
tion.

The field of the constructions of elementary geometry received at the hands of Apollonius an extension
in the solution of the problem to construct a circle
tangent to three given circles, and in the systematic
introduction of the diorismus (determination or limitation). This also appears in more difficult problems
in his *Conic Sections*, from which we see that Apollonius gives not simply the conditions for the possibility
of the solution in general, but especially desires to
determine the limits of the solutions.

From Zenodorus several theorems regarding isoperimetric figures are still extant; for example, he
states that the circle has a greater area than any isoperimetric regular polygon, that among all isoperimetric polygons of the same number of sides the regular has the greatest area, and so on. Hypsicles gives

*Chasles, *Aperçu historique sur l'origine et le développement des méthodes
en géométrie,* 1875. Hereafter referred to as Chasles.

as something new the division of the circumference
into three hundred and sixty degrees. From Heron
we have a book on geometry (according to Tannery
still another, a commentary on Euclid's Elements)
which deals in an extended manner with the mensuration of plane figures. Here we find deduced for the
area Δ of the triangle whose sides are a, b, and c,
where $2s = a + b + c$, the formula

$$\Delta = \sqrt{s(s-a)(s-b)(s-c)}.$$

In the measurement of the circle we usually find $\frac{22}{7}$ as
an approximation for π; but still in the *Book of Measurements* we also find $\pi = 3$.

In the period after the commencement of the
Christian era the output becomes still more meager.
Only occasionally do we find anything noteworthy.
Serenus, however, gives a theorem on transversals
which expresses the fact that a harmonic pencil is cut
by an arbitrary transversal in a harmonic range. In
the *Almagest* occurs the theorem regarding the inscribed quadrilateral, ordinarily known as Ptolemy's
Theorem, and a value written in sexagesimal form
$\pi = 3.8.30$, i. e.,

$$\pi = 3 + \frac{8}{60} + \frac{30}{60 \cdot 60} = 3\frac{17}{120} = 3.14166 \ldots . *$$

In a special treatise on geometry Ptolemy shows that
he does not regard Euclid's theory of parallels as indisputable.

*Cantor, I., p. 394.

To the last supporters of Greek geometry belong
Sextus Julius Africanus, who determined the width of
a stream by the use of similar right-angled triangles,
and Pappus, whose name has become very well known
by reason of his *Collection*. This work consisting originally of eight books, of which the first is wholly
lost and the second in great part, presents the sub-
stance of the mathematical writings of special repute
in the time of the author, and in some places adds
corollaries. Since his work was evidently composed
with great conscientiousness, it has become one of
the most trustworthy sources for the study of the
mathematical history of ancient times. The geometric part of the *Collection* contains among other things
discussions of the three different means between two
line-segments, isoperimetric figures, and tangency of
circles. It also discusses similarity in the case of circles; so far at least as to show that all lines which
join the ends of parallel radii of two circles, drawn in
the same or in opposite directions, intersect in a fixed
point of the line of centers.

The Greeks rendered important service not simply
in the field of elementary geometry: they are also the
creators of the theory of conic sections. And as in
the one the name of Euclid, so in the other the name
of Apollonius of Perga has been the signal for controversy. The theory of the curves of second order
does not begin with Apollonius any more than does
Euclidean geometry begin with Euclid; but what the

Elements signify for elementary geometry, the eight books of the *Conics* signify for the theory of lines of the second order. Only the first four books of the *Conic Sections* of Apollonius are preserved in the Greek text: the next three are known through Arabic translations: the eighth book has never been found and is given up for lost, though its contents have been restored by Halley from references in Pappus. The first book deals with the formation of conics by plane sections of circular cones, with conjugate diameters, and with axes and tangents. The second has especially to do with asymptotes. These Apollonius obtains by laying off on a tangent from the point of contact the half-length of the parallel diameter and joining its extremity to the center of the curve. The third book contains theorems on foci and secants, and the fourth upon the intersection of circles with conics and of conics with one another. With this the elementary treatment of conics by Apollonius closes. The following books contain special investigations in application of the methods developed in the first four books. Thus the fifth book deals with the maximum and minimum lines which can be drawn from a point to the conic, and also with the normals from a given point in the plane of the curve of the second order; the sixth with equal and similar conics; the seventh in a remarkable manner with the parallelograms having conjugate diameters as sides and the theorem upon the sum of the squares of conjugate diameters. The eighth book contained, according to Halley, a series of problems connected in the closest manner with lemmas of the seventh book.

The first effort toward the development of the theory of conic sections is ascribed to Hippocrates.[*] He reduced the duplication of the cube to the construction of two mean proportionals x and y between two given line-segments a and b; thus[†]

$$\frac{a}{x} = \frac{x}{y} = \frac{y}{b} \text{ gives } x^2 = ay,\ y^2 = bx, \text{ whence}$$

$$x^3 = a^2 b = \frac{b}{a} \cdot a^3 = m \cdot a^3.$$

Archytas and Eudoxus seem to have found, by plane construction, curves satisfying the above equations but different from straight lines and circles. Menaechmus sought for the new curves, already known by plane constructions, a representation by sections of cones of revolution, and became the discoverer of conic sections in this sense. He employed only sections perpendicular to an element of a right circular cone; thus the parabola was designated as the "section of a right-angled cone" (whose generating angle is 45°); the ellipse, the "section of an acute-angled cone"; the hyperbola, the "section of an obtuse-angled cone." These names are also used by Archimedes, although he was aware that the three curves can be formed as sections of any circular cone. Apol-

[*] Zeuthen, *Die Lehre von den Kegelschnitten im Altertum.* Deutsch von v. Fischer-Benzon, 1886. P. 459. Hereafter referred to as Zeuthen.

[†] Cantor, I., p. 200.

lonius first introduced the names "ellipse,' "parabola," "hyperbola." Possibly Menaechmus, but in any case Archimedes, determined conics by a linear equation between areas, of the form $y^2 = kxx_1$. The semi-parameter, with Archimedes and possibly some of his predecessors, was known as "the segment to the axis," i. e., the segment of the axis of the circle from the vertex of the curve to its intersection with the axis of the cone. The designation "parameter" is due to Desargues (1639).[*]

It has been shown[†] that Apollonius represented the conics by equations of the form $y^2 = px + ax^2$, where x and y are regarded as parallel coördinates and every term is represented as an area. From this other linear equations involving areas were derived, and so equations belonging to analytic geometry were obtained by the use of a system of parallel coördinates whose origin could, for geometric reasons, be shifted simultaneously with an interchange of axes. Hence we already find certain fundamental ideas of the analytic geometry which appeared almost two thousand years later.

The study of conic sections was continued upon the cone itself only till the time when a single fundamental plane property rendered it possible to undertake the further investigation in the plane.[‡] In this way there had become known, up to the time of Archimedes, a number of important theorems on conjugate diameters, and the relations of the lines to these diameters as axes, by the aid of linear equations be-

[*]Baltzer, R., *Analytische Geometrie*, 1882.
[†]Zeuthen, p. 32.   [‡]Zeuthen, p. 43.

tween areas. There were also known the so-called
Newton's power-theorem, the theorem that the rectangles of the segments of two secants of a conic drawn
through an arbitrary point in given direction are in a
constant ratio, theorems upon the generation of a
conic by aid of its tangents or as the locus related to
four straight lines, and the theorem regarding pole
and polar. But these theorems were always applied
to only one branch of the hyperbola. One of the valuable services of Apollonius was to extend his own
theorems, and consequently those already known, to
both branches of the hyperbola. His whole method
justifies us in regarding him the most prominent rep-
resentative of the Greek theory of conic sections, and
so much the more when we can see from his principal
work that the foundations for the theory of projective
ranges and pencils had virtually been laid by the ancients in different theorems and applications.

With Apollonius the period of new discoveries in
the realm of the theory of conics comes to an end. In
later times we find only applications of long known
theorems to problems of no great difficulty. Indeed,
the solution of problems already played an important
part in the oldest times of Greek geometry and furnished the occasion for the exposition not only of
conics but also of curves of higher order than the second. In the number of problems, which on account
of their classic value have been transmitted from generation to generation and have continually furnished

occasion for further investigation, three, by reason of
their importance, stand forth preëminent: the duplication of the cube, or more generally the multiplication
of the cube, the trisection of the angle and the quadrature of the circle. The appearance of these three
problems has been of the greatest significance in the
development of the whole of mathematics. The first
requires the solution of an equation of the third degree; the second (for certain angles at least) leads to
an important section of the theory of numbers, i. e.,
to the cyclotomic equations, and Gauss (see p. 160)
was the first to show that by a finite number of operations with straight edge and compasses we can construct a regular polygon of n sides only when $n-1$
$= 2^{2^p}$ (p an arbitrary integer). The third problem
reaches over into the province of algebra, for Lindemann[*] in the year 1882 showed that π cannot be the
root of an algebraic equation with integral coefficients.

The multiplication of the cube, algebraically the
determination of x from the equation

$$x^3 = \frac{b}{a} \cdot a^3 = m \cdot a^3,$$

is also called the Delian problem, because the Delians
were required to double their cubical altar.[†] The solution of this problem was specially studied by Plato,
Archytas, and Menaechmus; the latter solved it by

[*] *Mathem. Annalen*, XX., p. 215. See also *Mathem. Annalen*, XLIII., and
Klein, *Famous Problems of Elementary Geometry*, 1895, translated by Beman
and Smith, Boston, 1897.

[†] Cantor, I., p. 219.

the use of conics (hyperbolas and parabolas). Eratosthenes constructed a mechanical apparatus for the
same purpose.

Among the solutions of the problem of the trisection of an angle, the method of Archimedes is noteworthy. It furnishes an example of the so-called
"insertions" of which the Greeks made use when a
solution by straight edge and compasses was impossible. His process was as follows: Required to divide
the arc AB of the circle with center M into three
equal parts. Draw the diameter AE, and through B
a secant cutting the circumference in C and the diameter AE in D, so that CD equals the radius r of
the circle. Then arc $CE = \frac{1}{3}AB$.

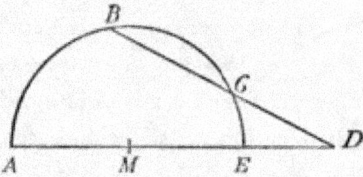

According to the rules of insertion the process consists in laying off upon a ruler a length r, causing it
to pass through B while one extremity D of the segment r slides along the diameter AE. By moving
the ruler we get a certain position in which the other
extremity of the segment r falls upon the circumference, and thus the point C is determined.

This problem Pappus claims to have solved after
the manner of the ancients by the use of conic sec-

tions. Since in the writings of Apollonius, so largely
lost, lines of the second order find an extended application to the solution of problems, the conics were
frequently called solid loci in opposition to plane loci,
i. e., the straight line and circle. Following these
came linear loci, a term including all other curves, of
which a large number were investigated.

This designation of the loci is found, for example,
in Pappus, who says in his seventh book* that a problem is called plane, solid, or linear, according as its
solution requires plane, solid, or linear loci. It is,
however, highly probable that the loci received their
names from problems, and that therefore the division
of problems into plane, solid, and linear preceded the
designation of the corresponding loci. First it is to
be noticed that we do not hear of "linear problems
and loci" till after the terms "plane and solid problems and loci" were in use. Plane problems were
those which in the geometric treatment proved to be
dependent upon equations of the first or second degree between segments, and hence could be solved
by the simple application of areas, the Greek method
for the solution of quadratic equations. Problems depending upon the solution of equations of the third
degree between segments led to the use of forms of
three dimensions, as, e. g., the duplication of the
cube, and were termed solid problems; the loci used
in their solution (the conics) were solid loci. At a

* Zeuthen, p. 205.

time when the significance of "plane" and "solid" was forgotten, the term "linear problem" was first applied to those problems whose treatment (by "linear loci") no longer led to equations of the first, second, and third degrees, and which therefore could no longer be represented as linear relations between segments, areas, or volumes.

Of linear loci Hippias applied the quadratrix (to which the name of Dinostratus was later attached through his attempt at the quadrature of the circle)[*] to the trisection of the angle. Eudoxus was acquainted with the sections of the torus made by planes parallel to the axis of the surface, especially the hippopede or figure-of-eight curve.[†] The spirals of Archimedes attained special celebrity. His exposition of their properties compares favorably with his elegant investigations of the quadrature of the parabola.

Conon had already generated the spiral of Archimedes[‡] by the motion of a point which recedes with uniform velocity along the radius OA of a circle k from the center O, while OA likewise revolves uniformly about O. But Archimedes was the first to discover certain of the beautiful properties of this curve; he found that if, after one revolution, the spiral meets the circle k of radius OA in B (where BO is tangent to the spiral at O), the area bounded by BO and the

* Cantor, I., pp. 184, 233.

† Majer, *Proklos über die Petita und Axiomata bei Euklid*, 1875.

‡ Cantor, I., p. 291.

spiral is one-third of the area of the circle k; further that the tangent to the spiral at B cuts off from a perpendicular to OB at O a segment equal to the circumference of the circle k.*

The only noteworthy discovery of Nicomedes is the construction of the conchoid which he employed to solve the problem of the two mean proportionals, or, what amounts to the same thing, the multiplication of the cube. The curve is the geometric locus of the point X upon a moving straight line g which constantly passes through a fixed point P and cuts a fixed straight line h in Y so that XY has a constant length. Nicomedes also investigated the properties of this curve and constructed an apparatus made of rulers for its mechanical description.

The cissoid of Diocles is also of use in the multiplication of the cube. It may be constructed as follows: Through the extremity A of the radius OA of a circle k passes the secant AC which cuts k in C and the radius OB perpendicular to OA in D; X, upon AC, is a point of the cissoid when $DX = DC$.† Geminus proves that besides the straight line and the circle the common helix invented by Archytas possesses the insertion property.

Along with the geometry of the plane was developed the geometry of space, first as elementary stere-

* Montucla.

† Klein, F., *Famous Problems of Elementary Geometry*, translated by Beman and Smith, Boston, 1897, p. 44.

ometry and then in theorems dealing with surfaces of
the second order. The knowledge of the five regular
bodies and the related circumscribed sphere certainly
goes back to Pythagoras. According to the statement
of Timaeus of Locri,* fire is made up of tetrahedra,
air of octahedra, water of icosahedra, earth of cubes,
while the dodecahedron forms the boundary of the
universe. Of these five cosmic or Platonic bodies
Theaetetus seems to have been the first to publish a
connected treatment. Eudoxus states that a pyramid
(or cone) is $\frac{1}{3}$ of a prism of equal base and altitude.
The eleventh, twelfth and thirteenth books of Euclid's
Elements offer a summary discussion of the ordinary
stereometry. (See p. 199.) Archimedes introduces
thirteen semi-regular solids, i. e., solids whose boundaries are regular polygons of two or three different
kinds. Besides this he compares the surface and volume of the sphere with the corresponding expressions
for the circumscribed cylinder and deduces theorems
which he esteems so highly that he expresses the desire to have the sphere and circumscribed cylinder
cut upon his tomb-stone. Among later mathematicians Hypsicles and Heron give exercises in the mensuration of regular and irregular solids. Pappus also
furnishes certain stereometric investigations of which
we specially mention as new only the determination
of the volume of a solid of revolution by means of the
meridian section and the path of its center of gravity.

* Cantor, I., p. 163.

He thus shows familiarity with a part of the theorem later known as Guldin's rule.

Of surfaces of the second order the Greeks knew the elementary surfaces of revolution, i. e., the sphere, the right circular cylinder and circular cone. Euclid deals only with cones of revolution, Archimedes on the contrary with circular cones in general. In addition, Archimedes investigates the "right-angled conoids" (paraboloids of revolution), the "obtuse-angled conoids" (hyperboloids of revolution of one sheet), and "long and flat spheroids" (ellipsoids of revolution about the major and minor axes). He determines the character of plane sections and the volume of segments of such surfaces. Probably Archimedes also knew that these surfaces form the geometric locus of a point whose distances from a fixed point and a given plane are in a constant ratio. According to Proclus,* who is of importance as a commentator upon Euclid, the torus was also known—a surface generated by a circle of radius r revolving about an axis in its plane so that its center describes a circle of radius e. The cases $r = e, > e, < e$ were discussed.

With methods of projection, also, the Greeks were not unacquainted.† Anaxagoras and Democritus are said to have known the laws of the vanishing point and of reduction, at least for the simplest cases. Hipparchus projects the celestial sphere from a pole upon

---

* Majer, *Proklos über die Petita und Axiomata bei Euklid*, 1875.

† Wiener.

the plane of the equator; he is therefore the inventor
of the stereographic projection which has come to be
known by the name of Ptolemy.

D. THIRD PERIOD.

ROMANS, HINDUS, CHINESE, ARABS.

Among no other people of antiquity did geometry
reach so high an eminence as among the Greeks.
Their acquisitions in this domain were in part transplanted to foreign soil, yet not so that (with the
possible exception of arithmetic calculation) anything
essentially new resulted. Frequently what was inherited from the Greeks was not even fully understood, and therefore remained buried in the literature
of the foreign nation. From the time of the Renaissance, however, but especially from that of Descartes,
an entirely new epoch with more powerful resources
investigated the ancient treasures and laid them under
contribution.

Among the Romans independent investigation of
mathematical truths almost wholly disappeared. What
they obtained from the Greeks was made to serve
practical ends exclusively. For this purpose parts of
Euclid and Heron were translated. To simplify the
work of the surveyors or agrimensores, important geometric theorems were collected into a larger work of
which fragments are preserved in the Codex Arceri-

anus. In the work of Vitruvius on architecture
($c.$—14) is found the value $\pi = 3\frac{1}{8}$ which, though less
accurate than Heron's value $\pi = 3\frac{1}{7}$, was more easily
employed in the duodecimal system.* Boethius has
left a special treatise on geometry, but the contents
are so paltry that it is safe to assume that he made
use of an earlier imperfect treatment of Greek geometry.

Although the Hindu geometry is dependent upon
the Greek, yet it has its own peculiarities due to the
arithmetical modes of thought of the people. Certain
parts of the Çulvasutras are geometric. These teach
the rope-stretching already known to the Egyptians,
i. e., they require the construction of a right angle by
means of a rope divided by a knot into segments 15
and 39 respectively, the ends being fastened to a segment 36 ($15^2 + 36^2 = 39^2$). They also use the gnomon
and deal with the transformation of figures and the
application of the Pythagorean theorem to the multiplication of a given square. Instead of the quadrature
of the circle appears the circulature of the square,†
i. e., the construction of a circle equal to a given
square. Here the diameter is put equal to $\frac{4}{5}$ of the
diagonal of the square, whence follows $\pi = 3\frac{1}{8}$ (the
value used among the Romans). In other cases a
process is carried on which yields the value $\pi = 3$.

The writings of Aryabhatta contain certain incor-
rect formulae for the mensuration of the pyramid and

*Cantor, I., p. 508.　　　†Cantor, I., p. 601.

sphere (for the pyramid $V=\frac{1}{3}Bh$), but also a number
of perfectly accurate geometric theorems. Aryabhatta
gives the approximate value $\pi=\frac{62832}{20000}=3.1416$.
Brahmagupta teaches mensurational or Heronic geometry and is familiar with the formula for the area
of the triangle,

$$\Delta=\sqrt{s\,(s-a)\,(s-b)\,(s-c)},$$

and the formula for the area of the inscribed quadrilateral,

$$i=\sqrt{(s-a)\,(s-b)\,(s-c)\,(s-d)},$$

which he applies incorrectly to any quadrilateral. In
his work besides $\pi=3$ we also find the value $\pi=\sqrt{10}$,
but without any indication as to how it was obtained.
Bhaskara likewise devotes himself only to algebraic
geometry.   For π he gives not only the Greek value
$\frac{22}{7}$ and that of Aryabhatta $\frac{62832}{20000}$, but also a value
$\pi=\frac{754}{240}=3.14166\ldots$ Of geometric demonstrations
Bhaskara knows nothing. He states the theorem,
adds the figure and writes "Behold!"*

In Bhaskara a transfer of geometry from Alexandria to India is undoubtedly demonstrable, and perhaps this influence extended still further eastward to
the Chinese.   In a Chinese work upon mathematics,
composed perhaps several centuries after Christ, the
Pythagorean theorem is applied to the triangle with
sides 3, 4, 5; rope-stretching is indicated; the ver

*Cantor, I., p. 614.

tices of a figure are designated by letters after the
Greek fashion; π is put equal to 3, and toward the
end of the sixth century to $\frac{22}{7}$.

Greek geometry reached the Arabs in part directly
and in part through the Hindus. The esteem, however, in which the classic works of Greek origin were
held could not make up for the lack of real productive power, and so the Arabs did not succeed in a
single point in carrying theoretic geometry, even in
the subject of conic sections, beyond what had been
reached in the golden age of Greek geometry. Only
a few particulars may be mentioned. In Al Khowarazmi is found a proof of the Pythagorean theorem
consisting only of the separation of a square into
eight isosceles right-angled triangles. On the whole
Al Khowarazmi draws more from Greek than from
Hindu sources. The classification of quadrilaterals
is that of Euclid; the calculations are made after
Heron's fashion. Besides the Greek value $\pi = \frac{22}{7}$ we
find the Hindu values $\pi = \frac{62832}{20000}$ and $\pi = \sqrt{10}$. Abul
Wafa wrote a book upon geometric constructions.
In this are found combinations of several squares into
a single one, as well as the construction of polyhedra
after the methods of Pappus. After the Greek fashion the trisection of the angle occupied the attention
of Tabit ibn Kurra, Al Kuhi, and Al Sagani. Among
later mathematicians the custom of reducing a geometric problem to the solution of an equation is common. It was thus that the Arabs by geometric solu-

tions attained some excellent results, but results of no
theoretic importance.

E. FOURTH PERIOD.

FROM GERBERT TO DESCARTES.

Among the Western nations we find the first traces
of geometry in the works of Gerbert, afterward known
as Pope Sylvester II.   Gerbert, as it seems, depends
upon the Codex Arcerianus, but also mentions Pythagoras and Eratosthenes.[*] We find scarcely anything
here besides field surveying as in Boethius.   Something more worthy first appears in Leonardo's (Fibonacci's) *Practica geometriae*[†] of 1220, in which work
reference is made to Euclid, Archimedes, Heron, and
Ptolemy.   The working over of the material handed
down from the ancients, in Leonardo's book, is fairly
independent.   Thus the rectification of the circle
shows where this mathematician, without making use
of Archimedes, determines from the regular polygon
of 96 sides the value $\pi = \dfrac{1440}{458\frac{1}{3}} = 3.1418$.

Since among the ancients no proper theory of star
polygons can be established, it is not to be wondered
at that the early Middle Ages have little to show in
this direction.   Star-polygons had first a mystic significance only ; they were used in the black art as the
pentacle, and also in architecture and heraldry.   Adel-

---

[*] Cantor, I., p. 810.   [†] Hankel, p. 344.

ard of Bath went with more detail into the study of star-polygons in his commentary on Euclidean geometry; the theory of these figures is first begun by Regiomontanus.

The first German mathematical work is the *Deut sche Sphära* written in Middle High German by Conrad of Megenberg, probably in Vienna in the first half of the fourteenth century. The first popular introduction to geometry appeared anonymously in the fifteenth century, in six leaves of simple rules of construction for geometric drawing. The beginning, containing the construction of BC perpendicular to AB by the aid of the right-angled triangle ABC in which BE bisects the hypotenuse AC, runs as follows:*
"From geometry some useful bits which are written after this. 1. First to make a right angle quickly. Draw two lines across each other just about as you wish and where the lines cross each other there put an e. Then place the compasses with one foot upon the point e, and open them out as far as you wish, and make upon each line a point. Let these be the letters a, b, c, all at one distance. Then make a line from a to b and from b to c. So you have a right angle of which here is an example."

This construction of a right angle, not given in Euclid but first in Proclus, appears about the year 1500 to be in much more extensive use than the method of Euclid by the aid of the angle inscribed in

* Günther, p. 347.

a semi-circle. By his knowledge of this last construction Adam Riese is said to have humiliated an architect who knew how to draw a right angle only by the
method of Proclus.

Very old printed works on geometry in German are *Dz Puech-
len der fialen gerechtikait* by Mathias Roriczer (1486) and Albrecht Dürer's *Underweysung der messung mit dem zirckel
und richtscheyt* (Nuremberg, 1525). The former gives in rather
unscientific manner rules for a special problem of Gothic architecture ; the latter, however, is a far more original work and on that
account possesses more interest.*

With the extension of geometric knowledge in
Germany Widmann and Stifel were especially concerned. Widmann's geometry, like the elements of
Euclid, begins with explanations: " Punctus is a small
thing that cannot be divided. Angulus is a corner
which is made there by two lines."† Quadrilaterals
have Arab names, a striking evidence that the ancient
Greek science was brought into the West by Arab influence. Nevertheless, by Roman writers (Boethius)
Widmann is led into many errors, as, e. g., when he
gives the area of the isosceles triangle of side a as $\frac{1}{2}a^2$.

In Rudolff's Coss, in the theory of powers, Stifel
has occasion to speak of a subject which first receives
proper estimation in the modern geometry, viz., the
right to admit more than three dimensions. "Since,
however, we are in arithmetic where it is permitted
to invent many things that otherwise have no form,

*Günther in *Schlömilch's Zeitschrift*, XX., Hl. 2.
† Gerhardt. *Geschichte der Mathematik in Deutschland*, 1877.

this also is permitted which geometry does not allow, namely to assume solid lines and surfaces and go beyond the cube just as if there were more than three dimensions, which is, of course, against nature. . . . But we have such good indulgence on account of the charming and wonderful usage of Coss."[*]

Stifel after the manner of Ptolemy extends the study of regular polygons and after the manner of Euclid the construction of regular solids. He discusses the quadrature of the circle, considering the latter as a polygon of infinitely many sides, and declares the quadrature impossible. According to Albrecht Dürer's *Underweysung*, etc., the quadrature of the circle is obtained when the diagonal of the square contains ten parts of which the diameter of the circle contains eight, i. e., $\pi = 3\frac{1}{8}$. It is expressly stated, however, that this is only an approximate construction. "We should need to know *quadratura circuli*, that is the making equal a circle and a square, so that the one should contain as much as the other, but this has not yet been demonstrated mechanically by scholars; but that is merely incidental; therefore so that in practice it may fail only slightly, if at all, they may be made equal as follows.[†]

[*] Stifel, *Die Coss Christoffs Rudolffs*. Mit schönen Exempeln der Coss. Durch Michael Stifel Gebessert vnd sehr gemehrt. . . . Gegeben zum Habersten | bei Königsberg in Preussen | den letzten tag dess Herbstmonds | im Jar 1552. . . . Zu Amsterdam Getruckt bey Wilhem Janson. Im Jar 1615.

[†] Dürer, *Underweysung der messung mit dem zirckel vnd richtscheyt in Linien ebnen vnd gantzen corporen*. Durch Albrecht Dürer zusamen getzogn vnd zu nutz alln kunstlieb habenden mit zu gebörigen figuren in truck gebracht im jar MDXXV. (Consists of *vier Büchlein*.)

Upon the mensuration of the circle[*] there appeared in 1584 a work by Simon van der Eycke in which the value $\pi = \dfrac{1521}{484}$ was given. By calculating the side of the regular polygon of 192 sides Ludolph van Ceulen found (probably in 1585) that $\pi < 3.14205 < \dfrac{1521}{484}$. In his reply Simon v. d. Eycke determined $\pi = 3.1446055$, whereupon L. v. Ceulen in 1586 computed π between 3.142732 and 3.14103. Ludolph van Ceulen's papers contain a value of π to 35 places, and this value of the Ludolphian number was put upon his tombstone (no longer known) in St. Peter's Church in Leyden. Ceulen's investigations led Snellius, Huygens, and others to further studies. By the theory of rapidly converging series it was first made possible to compute π to 500 and more decimals.[†]

A revival of geometry accompanied the activity of Vieta and Kepler. With these investigators begins a period in which the mathematical spirit commences to reach out beyond the works of the ancients.[‡] Vieta completes the analytic method of Plato; in an ingenious way he discusses the geometric construction of roots of equations of the second and third degrees; he also solves in an elementary manner the problem of the circle tangent to three given circles. Still more important results are secured by Kepler. For him geometry furnishes the key to the secrets of the world. With sure step he follows the path of induction and in his geometric investigations freely conforms to Euclid. Kepler established the symbolism of the "golden section," that problem of Eudoxus

---

[*] Rudio, F., *Das Problem von der Quadratur des Zirkels*, Zürich, 1890.

[†] D. Bierens de Haan in *Nieuw. Arch.*, I.; Cantor, II., p. 551.

[‡] Chasles.

stated in the sixth book of Euclid's *Elements*: "To divide a limited straight line in extreme and mean ratio."* This problem, for which Kepler introduced the designation *sectio divina* as well as *proportio divina*, is in his eyes of so great importance that he expresses himself: "Geometry has two great treasures: one is the theorem of Pythagoras, the other the division of a line in extreme and mean ratio. The first we may compare to a mass of gold, the second we may call a precious jewel."

The expression "golden section" is of more modern origin. It occurs in none of the text-books of the eighteenth century and appears to have been formed by a transfer from ordinary arithmetic. In the arithmetic of the sixteenth and seventeenth centuries the rule of three is frequently called the "golden rule." Since the beginning of the nineteenth century this golden rule has given way more and more before the so-called *Schlussrechnen* (analysis) of the Pestalozzi school. Consequently in place of the "golden rule," which is no longer known to the arithmetics, there appeared in the elementary geometries about the middle of the nineteenth century the "golden section," probably in connection with contemporary endeavors to attribute to this geometric construction the importance of a natural law.

Led on by his astronomical speculations, Kepler made a special study of regular polygons and star-polygons. He considered groups of regular polygons capable of elementary construction, viz., the series of polygons with the number of sides given by $4 \cdot 2^n$, $3 \cdot 2^n$, $5 \cdot 2^n$, $15 \cdot 2^n$ (from $n = 0$ on), and remarked that

* Sonnenburg, *Der goldene Schnitt*, 1881.

a regular heptagon cannot be constructed by the help
of the straight line and circle alone. Further there is
no doubt that Kepler well understood the *Conics* of
Apollonius and had experience in the solution of problems by the aid of these curves. In his works we
first find the term "foci" for those points of conic
sections which in earlier usage are known as *puncta
ex comparatione, puncta ex applicatione facta, umbilici,*
or "poles";* also the term "eccentricity" for the
distance from a focus to the center divided by the
semi-major axis, of the curve of the second order, and
the name "eccentric anomaly" for the angle $P'OA$,
where OA is the semi-major axis of an ellipse and P'
the point in which the ordinate of a point P on the
curve intersects the circle upon the major axis.†

Also in stereometric investigations, which had been
cultivated to a decided extent by Dürer and Stifel,
Kepler is preëminent among his contemporaries. In
his *Harmonice Mundi* he deals not simply with the
five regular Platonic and thirteen semi-regular Archi-
medean solids, but also with star-polygons and star-
dodecahedra of twelve and twenty vertices. Besides
this we find the determination of the volumes of solids
obtained by the revolution of conics about diameters,
tangents, or secants. Similar determinations of volumes were effected by Cavalieri and Guldin. The
former employed a happy modification of the method

* C. Taylor, in *Cambr. Proc.*, IV.
† Baltzer, R., *Analytische Geometrie*, 1882.

of exhaustions, the latter used a rule already known
to Pappus but not accurately established by him.

To this period belong the oldest known attempts
to solve geometric problems with only one opening of
the compasses, an endeavor which first found accurate
scientific expression in Steiner's *Geometrische Con-
struktionen, ausgeführt mittels der geraden Linie und
eines festen Kreises* (1833). The first traces of such
constructions go back to Abul Wafa.* From the Arabs
they were transmitted to the Italian school where they
appear in the works of Leonardo da Vinci and Cardan.
The latter received his impulse from Tartaglia who
used processes of this sort in his problem-duel with
Cardan and Ferrari. They also occur in the *Resolutio
omnium Euclidis problematum* (Venice, 1553) of Bene-
dictis, a pupil of Cardan, in the *Geometria deutsch* and
in the construction of a regular pentagon by Dürer.
In his *Underweysung*, etc., Dürer gives a geometrically
accurate construction of the regular pentagon but also
an approximate construction of the same figure to be
made with a circle of fixed radius.

This method of constructing a regular pentagon on *AB* is as
follows: About *A* and *B* as centers, with radius *AB*, construct circles intersecting in *C* and *D*. The circle about *D* as a center with
the same radius cuts the circles with centers at *A* and *B* in *E* and
F and the common chord *CD* in *G*. The same circles are cut by
FG and *EG* in *J* and *H*. *AJ* and *BH* are sides of the regular
pentagon. (The calculation of this symmetric pentagon shows

* Günther in *Schlömilch's Zeitschrift*, XX. Cantor, I., p. 700.

HBA = 108° 20′, while the corresponding angle of the regular pentagon is 108°.)

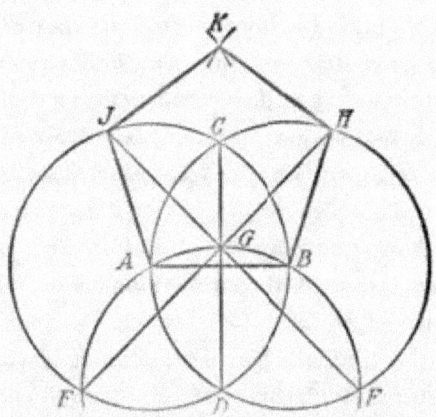

In Dürer and all his successors who write upon rules of geometric construction, we find an approximate construction of the regular heptagon : "The side of the regular heptagon is half that of the equilateral triangle," while from calculation the half side of the equilateral triangle = 0.998 of the side of the heptagon. Daniel Schwenter likewise gave constructions with a single opening of the compasses in his *Geometria practica nova et aucta* (1625). Dürer, as is manifest from his work *Underweysung der messung*, etc., already cited several times, also rendered decided service in the theory of higher curves. He gave a general conception of the notion of asymptotes and found as new forms of higher curves certain cyclic curves and mussel-shaped lines.

From the fifteenth century on, the methods of projection make a further advance. Jan van Eyck[*] in the great altar painting in Ghent makes use of the laws of perspective, e. g., in the application of the

* Wiener.

vanishing point, but without a mathematical grasp of these laws. This is first accomplished by Albrecht Dürer who in his *Underweysung der messung mit dem zirckel und richtscheyt* makes use of the point of sight and distance-point and shows how to construct the perspective picture from the ground plan and elevation. In Italy perspective was developed by the architect Brunelleschi and the sculptor Donatello. The first work upon this new theory is due to the architect Leo Battista Alberti. In this he explains the perspective image as the intersection of the pyramid of visual rays with the picture-plane. He also mentions an instrument for constructing it, which consists of a frame with a quadratic net-work of threads and a similar net-work of lines upon the drawing surface. He also gives the method of the distance-point as invented by him, by means of which he then pictures the ground divided into quadratic figures.* This process received a further extension at the hands of Piero della Francesca who employed the vanishing points of arbitrary horizontal lines.

In German territory perspective was cultivated with special zeal in Nuremberg where the goldsmith Lencker, some decades after Dürer, extended the latter's methods. The first French study of perspective is due to the artist J. Cousin (1560) who in his *Livre de la perspective* made use of the point of sight and the distance-point, besides the vanishing points of hori-

* Wiener.

zontal lines, after the manner of Piero. Guido Ubaldi
goes noticeably further when he introduces the vanishing point of series of parallel lines of arbitrary direction. What Ubaldi simply foreshadows, Simon
Stevin clearly grasps in its principal features, and in
an important theorem he lays the foundation for the
development of the theory of collineation.

F. FIFTH PERIOD.

FROM DESCARTES TO THE PRESENT.

Since the time of Apollonius many centuries had
elapsed and yet no one had succeeded in reaching the
full height of Greek geometry. This was partly because the sources of information were relatively few,
and attainable indirectly and with difficulty, and partly
because men, unfamiliar with Greek methods of investigation, looked upon them with devout astonishment. From this condition of partial paralysis, and
of helpless endeavor longing for relief, geometry was
delivered by Descartes. This was not by a simple addition of related ideas to the old geometry, but merely
by the union of algebra with geometry, thus giving
rise to analytic geometry.

By way of preparation many mathematicians, first
of all Apollonius, had referred the most important elementary curves, namely the conics, to their diameters
and tangents and had expressed this relation by equa-

tions of the first degree between areas, so that certain relations were obtained between line-segments identical with abscissas and ordinates.

In the conics of Apollonius we find expressions which have been translated "ordinatim applicatae" and "abscissae." For the former expression Fermat used "applicate" while others wrote "ordinate." Since the time of Leibnitz (1692) abscissas and ordinates have been called "co-ordinates."*

Even in the fourteenth century we find as an object of study in the universities a kind of co-ordinate geometry, the "latitudines formarum." "Latitudo"† signified the ordinate, "longitudo" the abscissa of a variable point referred to a system of rectangular co-ordinates, and the different positions of this point formed the "figura." The technical words longitude and latitude had evidently been borrowed from the language of astronomy. In practice of this art Oresme confined himself to the first quadrant in which he dealt with straight lines, circles, and even the parabola, but always so that only a positive value of a co-ordinate was considered.

Among the predecessors of Descartes we reckon, besides Apollonius, especially, Vieta, Oresme, Cavalieri, Roberval, and Fermat, the last the most distinguished in this field; but nowhere, even by Fermat, had any attempt been made to refer several curves of

* Baltzer, R., *Analytische Geometrie*, 1882.
† Günther, p. 181.

different orders simultaneously to one system of co-
ordinates, which at most possessed special significance
for one of the curves.   It is exactly this thing which
Descartes systematically accomplished.

The thought with which Descartes made the laws
of arithmetic subservient to geometry is set forth by
himself in the following manner: *

"All problems of geometry may be reduced to such
terms that for their construction we need only to know
the length of certain right lines.   And just as arithmetic as a whole comprises only four or five operations, viz., addition, subtraction, multiplication, division, and evolution, which may be considered as a
kind of division, so in geometry to prepare the lines
sought to be known we have only to add other lines
to them or subtract others from them ; or, having one
which I call unity (so as better to refer it to numbers),
which can ordinarily be taken at pleasure, having two
others to find a fourth which shall be to one of these
as the other is to unity, which is the same as multiplication ;† or to find a fourth which shall be to one
of the two as unity is to the other which is the same
as division ;‡ or finally to find one or two or several
mean proportionals between unity and any other line,
which is the same as to extract the square, cube, . . .
root.§   I shall not hesitate to introduce these terms

* Marie, M., *Histoire des Sciences Mathématiques et Physiques*, 1883-1887.

† $c:a=b:1,\ c=ab$.

‡ $c:a=1:b,\ c=a:b$.

§ $1:a=a:b=b:c=c:d=\ldots$ gives $a=\sqrt{b}=\sqrt[3]{c}=\sqrt[4]{d}$..

of arithmetic into geometry in order to render myself more intelligible. It should be observed that, by a^2, b^3, and similar quantities, I understand as usual simple lines, and that I call them square or cube only so as to employ the ordinary terms of algebra." (a^2 is the third proportional to unity and a, or $1 : a = a : a^2$, and similarly $b : b^2 = b^2 : b^3$.)

This method of considering arithmetical expressions was especially influenced by the geometric discoveries of Descartes. As Apollonius had already determined points of a conic section by parallel chords, together with the distances from a tangent belonging to the same system, measured in the direction of the conjugate diameter, so with Descartes every point of a curve is the intersection of two straight lines. Apollonius and all his successors, however, apply such systems of parallel lines only occasionally and that for the sole purpose of presenting some definite property of the conics with especial distinctness. Descartes, on the contrary, separates these systems of parallel lines from the curves, assigns them an independent existence and so obtains for every point on the curve a relation between two segments of given direction, which is nothing else than an equation. The geometric study of the properties of this curve can then be replaced by the discussion of the equation after the methods of algebra. The fundamental elements for the determination of a point of a curve are its co-ordinates, and from long known theorems it was evident

that a point of the plane can be fixed by two co-ordinates, a point of space by three.

Descartes's *Geometry* is not, perhaps, a treatise
on analytic geometry, but only a brief sketch which
sets forth the foundations of this theory in outline.
Of the three books which constitute the whole work
only the first two deal with geometry; the third is of
algebraic nature and contains the celebrated rule of
signs illustrated by a simple example, as well as the
solution of equations of the third and fourth degrees
with the construction of their roots by the use of
conics.

The first impulse to his geometric reflections was
due, as Descartes himself says, to a problem which
according to Pappus had already occupied the attention of Euclid and Apollonius. It is the problem to
find a certain locus related to three, four, or several
lines. Denoting the distances, measured in given directions, of a point P from the straight lines $g_1, g_2 \ldots$
g_n by $c_1, c_2 \ldots c_n$, respectively, we shall have

$$\text{for three straight lines}: \frac{c_1 c_2}{a c_3} = k,$$

$$\text{for four straight lines}: \frac{c_1 c_2}{c_3 c_4} = k,$$

$$\text{for five straight lines}: \frac{c_1 c_2 c_3}{a c_4 c_5} = k,$$

and so on. The Greeks originated the solution of the
first two cases, which furnish conic sections. No example could have shown better the advantage of the

new method. For the case of three lines Descartes denotes a distance by y, the segment of the corresponding line between the foot of this perpendicular and a fixed point by x, and shows that every other segment involved in the problem can be easily constructed. Further he states "that if we allow y to grow gradually by infinitesimal increments, x will grow in the same way and thus we may get infinitely many points of the locus in question."

The curves with which Descartes makes us gradually familiar he classifies so that lines of the first and second orders form a first group, those of the third and fourth orders a second, those of the fifth and sixth orders a third, and so on. Newton was the first to call a curve, which is defined by an algebraic equation of the nth degree between parallel co-ordinates, a line of the nth order, or a curve of the $(n—1)$th class. The division into algebraic and transcendental curves was introduced by Leibnitz; previously, after the Greek fashion, the former had been called geometric, the latter mechanical lines.*

Among the applications which Descartes makes, the problem of tangents is prominent. This he treats in a peculiar way: Having drawn a normal to a curve at the point P, he describes a circle through P with the center at the intersection of this normal with the

* Baltzer, R., *Analytische Geometrie*, 1882. Up to the time of Descartes all lines except straight lines and conics were called mechanical. He was the first to apply the term geometric lines to curves of degree higher than the second.

X-axis, and asserts that this circle cuts the curve at P in two consecutive points; i. e., he states the condition that after the elimination of x the equation in y shall have a double root.

A natural consequence of the acceptance of the Cartesian co-ordinate system was the admission of negative roots of algebraic equations. These negative roots had now a real significance; they could be represented, and hence were entitled to the same rights as positive roots.

In the period immediately following Descartes, geometry was enriched by the labors of Cavalieri, Fermat, Roberval, Wallis, Pascal, and Newton, not at first by a simple application of the co-ordinate geometry, but often after the manner of the ancient Greek geometry, though with some of the methods essentially improved. The latter is especially true of Cavalieri, the inventor of the method of indivisibles,* which a little later was displaced by the integral ca`culus, but may find a place here since it rendered service to geometry exclusively. Cavalieri enjoyed working with the geometry of the ancients. For example, he was the first to give a satisfactory proof of the so-called Guldin's rule already stated by Pappus. His chief endeavor was to find a general process for the determination of areas and volumes as well as centers of gravity, and for this purpose he remodelled the

*In French works *Méthode des indivisibles*, originally in the work *Geometria indivisibilibus continuorum nova quadam ratione promota*, Bologna, 1635.

method of exhaustions. Inasmuch as Cavalieri's method, of which he was master as early as 1629, may even to-day replace to advantage ordinary integration in elementary cases, its essential character may be set forth in brief outline.[*]

If $y = f(x)$ is the equation of a curve in rectangular co-ordinates, and he wishes to determine the area bounded by the axis of x, a portion of the curve, and the ordinates corresponding to x_0 and x_1, Cavalieri divides the difference $x_1 - x_0$ into n equal parts. Let h represent such a part and let n be taken very large. An element of the surface is then $= hy = hf(x)$, and the whole surface becomes

$$\sum_0^{n-1} h \cdot f(x_0 + nh).$$

For $n = \infty$ we evidently get exactly

$$\int_{x_0}^{x_1} f(x)\, dx.$$

But this is not the quantity which Cavalieri seeks to determine. He forms only the ratios of portions of the area sought, to the rectangle with base $x_1 - x_0$ and altitude y_1, so that the quantity to be determined is the following:

$$\frac{\sum_0^{n-1} h \cdot f(x_0 + nh)}{n \cdot h \cdot f(x_1)} = \frac{\sum_0^{n-1} f(x_0 + nh)}{n f(x_1)}$$

Cavalieri applies this formula, which he derives in

* Marie.

complete generality from grounds of analogy, only to
the case where $f(x)$ is of the form Ax^m $(m=2, 3, 4)$.
The extension to further cases was made by Rober
val, Wallis, and Pascal.

In the simplest cases the method of indivisibles gives the following results.[*] For a parallelogram the indivisible quantity or
element of surface is a parallel to the base; the number of indivisible quantities is proportional to the altitude; hence we have
as the measure of the area of the parallelogram the product of the
measures of the base and altitude. The corresponding conclusion
holds for the prism. In order to compare the area of a triangle
with that of the parallelogram of the same base and altitude, we
decompose each into elements by equidistant parallels to the base.
The elements of the triangle are then, beginning with the least, 1,
2, 3, . . . n; those of the parallelogram, n, n, . . . n. Hence the
ratio

$$\frac{\text{Triangle}}{\text{Parallelogram}} = \frac{1+2+\ldots+n}{n\cdot n} = \frac{\frac{1}{2}n(n+1)}{n^2} = \frac{1}{2}\left(1+\frac{1}{n}\right);$$

whence for $n=\infty$ we get the value $\frac{1}{2}$. For the corresponding solids
we get likewise

$$\frac{\text{Pyramid}}{\text{Prism}} = \frac{1^2+2^2+\ldots+n^2}{n^3} = \frac{\frac{1}{6}n(n+1)(2n+1)}{n^3}$$

$$= \frac{1}{6}\left(1+\frac{1}{n}\right)\left(2+\frac{1}{n}\right)\Big|_{n=\infty} = \frac{1}{3}.$$

After the lapse of a few decades this analytic-
geometric method of Cavalieri's was forced into the
background by the integral calculus, which could be
directly applied in all cases. At first, however, Roberval, known by his method of tangents, trod in the
footsteps of Cavalieri. Wallis used the works of Des-

* Marie.

cartes and Cavalieri simultaneously, and considered especially curves whose equations were of the form $y = x^m$, m integral or fractional, positive or negative. His chief service consists in this, that in his brilliant work he put a proper estimate upon Descartes's discovery and rendered it more accessible. In this work Wallis also defines the conics as curves of the second degree, a thing never before done in this definite manner.

Pascal proved to be a talented disciple of Cavalieri and Desargues. In his work on conics, composed about 1639 but now lost (save for a fragment),* we find Pascal's theorem of the inscribed hexagon or *Hexagramma mysticum* as he termed it, which Bessel rediscovered in 1820 without being aware of Pascal's earlier work,† also the theorem due to Desargues that if a straight line cuts a conic in P and Q, and the sides of an inscribed quadrilateral in A, B, C, D, we have the following equation :

$$\frac{PA \cdot PC}{PB \cdot PD} = \frac{QA \cdot QC}{QC \cdot QD}.$$

Pascal's last work deals with a curve called by him the roulette, by Roberval the trochoid, and generally known later as the cycloid. Bouvelles (1503) already knew the construction of this curve, as did Cardinal von Cusa in the preceding century.‡ Galileo, as is shown by a letter to Torricelli in 1639, had made (be

* Cantor, II., p. 622.   † Bianco in *Torino Att.*, XXI.
‡ Cantor, II., pp. 186, 351.

ginning in 1590) an exhaustive study of rolling curves
in connection with the construction of bridge arches.
The quadrature of the cycloid and the determination
of the volume obtained by revolution about its axis
had been effected by Roberval, and the construction
of the tangent by Descartes. In the year 1658 Pascal
was able to determine the length of an arc of a cy-
cloidal segment, the center of gravity of this surface,
and the corresponding solid of revolution. Later the
cycloid appears in physics as the brachistochrone and
tautochrone, since it permits a body sliding upon it to
pass from one fixed point to another in the shortest
time, while it brings a material point oscillating upon
it to its lowest position always in the same time.
Jacob and John Bernoulli, among others, gave attention to isoperimetric problems; but only the former
secured any results of value, by furnishing a rigid
method for their solution which received merely an
unimportant simplification from John Bernoulli. (See
pages 178–179.)

The decades following Pascal's activity were in
large part devoted to the study of tangent problems
and the allied normal problems, but at the same time
the general theory of plane curves was constantly
developing. Barrow gave a new method of determin
ing tangents, and Huygens studied evolutes of curves
and indicated the way of determining radii of curva
ture. From the consideration of caustics, Tschirn
hausen was led to involutes and Maclaurin constructed

the circle of curvature at any point of an algebraic curve. The most important extension of this theory was made in Newton's *Enumeratio linearum tertii ordinis* (1706). This treatise establishes the distinction between algebraic and transcendental curves. It then makes an exhaustive study of the equation of a curve of the third order, and thus finds numerous such curves which may be represented as "shadows" of five types, a result which involves an analytic theory of perspective. Newton knew how to construct conics from five tangents. He came upon this discovery in his endeavor to investigate "after the manner of the ancients" without analytic geometry. Further he considered multiple points of a curve at a finite distance and at infinity, and gave rules for investigating the course of a curve in the neighborhood of one of its points ("Newton's parallelogram" or "analytic triangle"), as also for the determination of the order of contact of two curves at one of their common points. (Leibnitz and Jacob Bernoulli had also written upon osculations; Plücker (1831) called the situation where two curves have k consecutive points in common "a k-pointic contact"; in the same case Lagrange (1779) had spoken of a "contact of $(k-1)$th order.")[†]

Additional work was done by Newton's disciples, Cotes and Maclaurin, as well as by Waring. Maclaurin made interesting investigations upon corre-

[*] Baltzer.
[†] Cayley, A., *Address to the British Association*, etc., 1883.

sponding points of a curve of the third order, and
thus showed that the theory of these curves was much
more comprehensive than that of conics. Euler likewise entered upon these investigations in his paper
*Sur une contradiction apparente dans la théorie des courbes
planes* (Berlin, 1748), where it is shown that by eight intersections of two curves of the third order the ninth is
completely determined. This theorem, which includes
Pascal's theorem for conics, introduced point groups,
or systems of points of intersection of two curves, into
geometry. This theorem of Euler's was noticed in
1750 by Cramer who gave special attention to the singularities of curves in his works upon the intersection
of two algebraic curves of higher order; hence the
obvious contradiction between the number of points
determining a plane curve and the number of independent intersections of two curves of the same order
bears the name of "Cramer's paradox." This contradiction was solved by Lamé in 1818 by the principle
which bears his name.* Partly in connection with
known results of the Greek geometry, and partly independently, the properties of certain algebraic and
transcendental curves were investigated. A curve
which is formed like the conchoid of Nicomedes, if
we replace the straight line by a circle, is called by

*Loria, G., *Die hauptsächlichsten Theorien der Geometrie in ihrer frühe-
ren und jetzigen Entwicklung*. Deutsch von Schütte, 1888. For a more accurate account of Cramer's paradox, in which proper credit is given to Maclaurin's discovery, see Scott, C. A., "On the Intersections of Plane Curves,"
Bull. Am. Math. Soc., March, 1898.

Roberval the limaçon of Pascal. The **cardioid** of the eighteenth century is a special case of this spiral. If, with reference to two fixed points A, B, a point P satisfies the condition that a linear function of the distances PA, PB has a constant value, then is the locus of P a Cartesian oval. This curve was found by Descartes in his studies in dioptrics. For $PA \cdot PB =$ constant, we have Cassini's oval, which the astronomer of Louis XIV. wished to regard as the orbit of a planet instead of Kepler's ellipse. In special cases **Cassini's** oval contains a loop, and this form received from Jacob Bernoulli (1694) the name lemniscate. With the investigation of the logarithmic curve $y = a^x$ was connected the study made by Jacob and John Bernoulli, Leibnitz, Huygens, and others, of the curve of equilibrium of an inextensible, flexible thread. This furnished the catenary (*catenaria*, 1691), the idea of which had already occurred to Galileo.* The group of spirals found by Archimedes was enlarged in the seventeenth and eighteenth centuries by the addition of the hyperbolic, parabolic, and logarithmic spirals, and Cotes's lituus (1722). In 1687 Tschirnhausen defined a quadratrix, differing from that of the Greeks, as the locus of a point P, lying at the same time upon $LQ \parallel BO$ and upon $MP \parallel OA$ (OAB is a quadrant), where L moves over the quadrant and M over the radius OB uniformly. Whole systems of curves and surfaces were considered. Here belong the investiga-

* Cantor, III., p. 211.

tions of involutes and evolutes, envelopes in general, due to Huygens, Tschirnhausen, John Bernoulli, Leibnitz, and others. The consideration of the pencil of rays through a point in the plane, and of the pencil of planes through a straight line in space, was introduced by Desargues, 1639.[*]

The extension of the Cartesian co-ordinate method to space of three dimensions was effected by the labors of Van Schooten, Parent, and Clairaut.[†] Parent represented a surface by an equation involving the three co-ordinates of a point in space, and Clairaut perfected this new procedure in a most essential manner by a classic work upon curves of double curvature. Scarcely thirty years later Euler established the analytic theory of the curvature of surfaces, and the classification of surfaces in accordance with theorems analogous to those used in plane geometry. He gives formulae of transformation of space co-ordinates and a discussion of the general equation of surfaces of the second order, with their classification. Instead of Euler's names: "elliptoid, elliptic-hyperbolic, hyperbolic-hyperbolic, elliptic-parabolic, parabolic-hyperbolic surface," the terms now in use, "ellipsoid, hyperboloid, paraboloid," were naturalized by Biot and Lacroix.[‡]

Certain special investigations are worthy of mention. In 1663 Wallis studied plane sections and effected the cubature, of a conoid with horizontal di-

* Baltzer.   † Loria.   ‡ Baltzer.

recting plane whose generatrix intersects a vertical directing straight line and vertical directing circle (*cono-cuneus*). To Wren we owe an investigation of the hyperboloid of revolution of two sheets (1669) which he called "cylindroid." The domain of gauche curves, of which the Greeks knew the common helix of Archytas and the spherical spiral corresponding in formation to the plane spiral of Archimedes, found an extension in the line which cuts under a constant angle the meridians of a sphere. Nuñez (1546) had recognized this curve as not plane, and Snellius (1624) had given it the name *loxodromia sphaerica*. The problem of the shortest line between two points of a surface, leading to gauche curves which the nineteenth century has termed "geodetic lines," was stated by John Bernoulli (1698) and taken in hand by him with good results. In a work of Pitot in 1724 (printed in 1726)* upon the helix, we find for the first time the expression *ligne à double courbure*, line of double curvature, for a gauche curve. In 1776 and 1780 Meusnier gave theorems upon the tangent planes to ruled surfaces, and upon the curvature of a surface at one of its points, as a preparation for the powerful development of the theory of surfaces soon to begin.†

There are still some minor investigations belonging to this period deserving of mention. The algebraic expression for the distance between the centers of the inscribed and circumscribed circles of a triangle

*Cantor, III., p. 428.   † Baltzer.

was determined by William Chapple (about 1746), afterwards by Landen (1755) and Euler (1765).* In 1769 Meister calculated the areas of polygons whose sides, limited by every two consecutive vertices, intersect so that the perimeter contains a certain number of double points and the polygon breaks up into cells with simple or multiple positive or negative areas. Upon the areas of such singular polygons Möbius published later investigations (1827 and 1865).* Saurin considered the tangents of a curve at multiple points and Ceva starting from static theorems studied the transversals of geometric figures. Stewart still further extended the theorems of Ceva, while Cotes determined the harmonic mean between the segments of a secant to a curve of the nth order reckoned from a fixed point. Carnot also extended the theory of transversals. Lhuilier solved the problem: In a circle to inscribe a polygon of n sides passing through n fixed points. Brianchon gave the theorem concerning the hexagon circumscribed about a conic dualistically related to Pascal's theorem upon the inscribed hexagon. The application of these two theorems to the surface of the sphere was effected by Hesse and Thieme. In the work of Hesse a Pascal hexagon is formed upon the sphere by six points which lie upon the intersection of the sphere with a cone of the second order having its vertex at the center of the sphere. Thieme selects a right circular cone. The material usually taken for the elementary geometry of the schools has among other things received an extension through numerous theorems upon the circle named after K. W. Feuerbach (1822), upon symmedian lines of a triangle, upon the Grebe point and the Brocard figures (discovered in part by Crelle, 1816; again introduced by Brocard, 1875).*

The theory of regular geometric figures received its most important extension at the hands of Gauss, who discovered noteworthy theorems upon the possibility or impossibility of elementary constructions of regular polygons. (See p. 160.) Poinsot elaborated the theory of the regular polyhedra by publishing his views on the five Platonic bodies and especially upon the "Kepler-Poinsot regular solids of higher class," viz., the four star-polyhedra which are formed from the icosahedron and dodecahedron. These studies were continued by Wiener, Hessel, and Hess, with the removal of certain restrictions, so that a whole series of solids, which in an extended sense may be regarded as regular, may be added to those named above. Corresponding studies for four-dimensional space have been undertaken by Scheffler, Rudel, Stringham, Hoppe, and Schlegel. They have determined that in such a space there exist six regular figures of which the simplest has as its boundary five tetrahedra. The boundaries of the remaining five fig-

* Lieber, *Ueber die Gegenmittellinie, den Grebe'schen Punkt und den Brocard'schen Kreis.* 1886–1888.

ures require 16 or 600 tetrahedra, 8 hexahedra, 24 octahedra, 120 dodecahedra.* It may be mentioned
further that in 1849 the prismatoid was introduced
into stereometry by E. F. August, and that Schubert
and Stoll so generalised the Apollonian contact problem as to be able to give the construction of the six-
teen spheres tangent to four given spheres.

Projective geometry, called less precisely modern
geometry or geometry of position, is essentially a
creation of the nineteenth century. The analytic geometry of Descartes, in connection with the higher
analysis created by Leibnitz and Newton, had regis-
tered a series of important discoveries in the domain
of the geometry of space, but it had not succeeded in
obtaining a satisfactory proof for theorems of pure
geometry. Relations of a specific geometric character
had, however, been discovered in constructive drawing. Newton's establishment of his five principal
types of curves of the third order, of which the sixty-
four remaining types may be regarded as projections,
had also given an impulse in the same direction. Still
more important were the preliminary works of Carnot,
which paved the way for the development of the new
theory by Poncelet, Chasles, Steiner, and von Staudt.
They it was who discovered ''the overflowing spring
of deep and elegant theorems which with astonishing
facility united into an organic whole, into the graceful
edifice of projective geometry, which, especially with

* Serret, *Essai d'une nouvelle méthode*, etc., 1873.

reference to the theory of curves of the second order, may be regarded as the ideal of a scientific organism."[*]

Projective geometry found its earliest unfolding on French soil in the *Géométrie descriptive* of Monge whose astonishing power of imagination, supported by the methods of descriptive geometry, discovered a host of properties of surfaces and curves applicable to the classification of figures in space. His work created "for geometry the hitherto unknown idea of geometric generality and geometric elegance,"[†] and the importance of his works is fundamental not only for the theory of projectivity but also for the theory of the curvature of surfaces. To the introduction of the imaginary into the considerations of pure geometry Monge likewise gave the first impulse, while his pupil Gaultier extended these investigations by defining the radical axis of two circles as a secant of the same passing through their intersections, whether real or imaginary.

The results of Monge's school thus derived, which were more closely related to pure geometry than to the analytic geometry of Descartes, consisted chiefly in a series of new and interesting theorems upon surfaces of the second order, and thus belonged to the same field that had been entered upon before Monge's time by Wren (1669), Parent and Euler. That Monge

* Brill, A., *Antrittsrede in Tübingen*, 1884.

† Hankel, *Die Elemente der projektivischen Geometrie in synthetischer Behandlung*, 1875.

did not hold analytic methods in light esteem is shown
by his *Application de l'algèbre à la géométrie* (1805) in
which, as Plücker says, "he introduced the equation
of the straight line into analytic geometry, thus laying
the foundation for the banishment of all constructions
from it, and gave it that new form which rendered
further extension possible."

While Monge was working by preference in the
space of three dimensions, Carnot was making a special study of ratios of magnitudes in figures cut by
transversals, and thus, by the introduction of the negative, was laying the foundation for a *géométrie de position* which, however, is not identical with the *Geometrie
der Lage* of to-day. Not the most important, but the
most noteworthy contribution for elementary school
geometry is that of Carnot's upon the complete quadrilateral and quadrangle.

Monge and Carnot having removed the obstacles
which stood in the way of a natural development of
geometry upon its own territory, these new ideas could
now be certain of a rapid development in well-pre-
pared soil. Poncelet furnished the seed. His work,
Traité des propriétés projectives des figures, which appeared in 1822, investigates those properties of figures
which remain unchanged in projection, i. e., their invariant properties. The projection is not made here,
as in Monge, by parallel rays in a given direction, but
by central projection, and so after the manner of perspective. In this way Poncelet came to introduce

the axis of perspective and center of perspective (according to Chasles, axis and center of homology) in the consideration of plane figures for which Desargues had already established the fundamental theorems. In 1811 Servois had used the expression "pole of a straight line," and in 1813 Gergonne the terms "polar of a point" and "duality," but in 1818 Poncelet developed some observations made by Lahire in 1685, upon the mutual correspondence of pole and polar in the case of conics, into a method of transforming figures into their reciprocal polars. Gergonne recognized in this theory of reciprocal polars a principle whose beginnings were known to Vieta, Lansberg, and Snellius, from spherical geometry. He called it the "principle of duality" (1826). In 1827 Gergonne associated dualistically with the notion of order of a plane curve that of its class. The line is of the nth order when a straight line of the plane cuts it in n points, of the nth class when from a point in the plane n tangents can be drawn to it.*

While in France Chasles alone interested himself thoroughly in its advancement, this new theory found its richest development in the third decade of the nineteenth century upon German soil, where almost at the same time the three great investigators, Möbius, Plücker, and Steiner entered the field. From this time on the synthetic and more constructive tendency followed by Steiner, von Staudt, and Möbius diverges†

---

* Baltzer.   † Brill, A., *Antrittsrede in Tübingen*, 1884.

from the analytic side of the modern geometry which Plücker, Hesse, Aronhold, and Clebsch had especially developed.

The *Barycentrischer Calcül* in the year 1827 furnished the first example of homogeneous co-ordinates, and along with them a symmetry in the developed formulae hitherto unknown to analytic geometry. In this calculus Möbius started with the assumption that every point in the plane of a triangle ABC may be regarded as the center of gravity of the triangle. In this case there belong to the points corresponding weights which are exactly the homogeneous co-ordinates of the point P with respect to the vertices of the fundamental triangle ABC. By means of this algorism Möbius found by algebraic methods a series of geometric theorems, for example those expressing invariant properties like the theorems on cross-ratios. These theorems, found analytically, Möbius sought to demonstrate geometrically also, and for this purpose he introduced with all its consequences the "law of signs" which expresses that for A, B, C, points of a straight line, $AB = -BA$, $AB + BA = 0$, $AB + BC + CA = 0$.

Independently of Möbius, but starting from the same principles, Bellavitis came upon his new geometric method of equipollences.* Two equal and parallel lines drawn in the same direction, AB and CD, are called equipollent (in Cayley's notation $AB \equiv CD$). By this assumption the whole theory is reduced to the

* Bellavitis, " Saggio di Applicazioni di un Nuovo Metodo di Geometria Analitica (Calcolo delle Equipollenze)," in *Ann. Lomb. Veneto*, t. 5, 1835.

consideration of segments proceeding from a fixed point. Further it is assumed that $AB + BC \equiv AC$ (Addition). Finally for the segments a, b, c, d, with inclinations a, β, γ, δ to a fixed axis, the equation $a \equiv \dfrac{bc}{d}$ must not only be a relation between lengths but must also show that $a = \beta + \gamma - \delta$ (Proportion). For $d = 1$ and $a = 0$ this becomes $a \equiv bc$, i. e., the product of the absolute values of the lengths is $a = bc$ and at the same time $a = \beta + \gamma$ (Multiplication). Equipollence is therefore only a special case of the equality of two objects, applied to segments.*

Möbius further introduced the consideration of correspondences of two geometric figures. The one-to-one correspondence, in which to every point of a first figure there corresponds one and only one point of a second figure and to every point of the second one and only one point of the first, Möbius called collineation. He constructed not only a collinear image of the plane but also of ordinary space.

These new and fundamental ideas which Möbius had laid down in the barycentric calculus remained for a long time almost unheeded and hence did not at once enter into the formation of geometric conceptions. The works of Plücker and Steiner found a more favorable soil. The latter "had recognized in immediate geometric perception the sufficient means and the only object of his knowledge. Plücker, on the other hand,† sought his proofs in the identity of the analytic operation and the geometric construc-

---

* Stolz, O., *Vorlesungen über allgemeine Arithmetik*, 1885–1886.

†"Clebsch, Versuch einer Darlegung und Würdigung seiner wissenschaftlichen Leistungen von einigen seiner Freunde (Brill, Gordan, Klein, Lüroth, A. Mayer, Nöther, Von der Mühll)," in *Math. Ann.*, Bd. 7.

tion, and regarded geometric truth only as one of the
many conceivable antitypes of analytic relation."

At a later period (1855) Möbius engaged in the
study of involutions of higher degree.   Such an involution of the mth degree consists of two groups each
of m points: A_1, A_2, A_3, . . . A_m; B_1, B_2, B_3, . . . B_m,
which form two figures in such a way that to the 1st,
2d, 3d, . . . mth points of one group, as points of the
first figure, there correspond in succession the 2d, 3d,
4th . . . 1st points of the same group as points of the
second figure, with the same determinate relation. Involutions of higher degree had been previously studied
by Poncelet (1843).   He started from the theorem
given by Sturm (1826), that by the conic sections of
the surfaces of the second order $u = 0$, $v = 0$, $u + \lambda v$
$= 0$, there are determined upon a straight line six
points, A, A', B, B', C, C' in involution, i. e., so that
in the systems $ABCA'B'C'$ and $A'B'C'ABC$ not only
A and A', B and B', C and C', but also A' and A, B'
and B, C' and C are corresponding point-pairs. This
mutual correspondence of three point-pairs of a line
Desargues had already (in 1639) designated by the
term "involution." [*]

Plücker is the real founder of the modern analytic
tendency, and he attained this distinction by "formu-
lating analytically the principle of duality and following out its consequences." [†]   His *Analytisch-geometrische Untersuchungen* appeared in 1828. By this work

[*] Baltzer.   [†] Brill, A., *Antrittsrede in Tübingen*, 1884.

was created for geometry the method of symbolic notation and of undetermined coefficients, whereby one
is freed from the necessity, in the consideration of the
mutual relations of two figures, of referring to the
system of co-ordinates, so that he can deal with the
figures themselves. The *System der analytischen Geometrie* of 1835 furnishes, besides the abundant application of the abbreviated notation, a complete classification of plane curves of the third order. In the
Theorie der algebraischen Kurven of 1839, in addition
to an investigation of plane curves of the fourth order
there appeared those analytic relations between the
ordinary singularities of plane curves which are generally known as "Plücker's equations."

These Plücker equations which at first are applied
only to the four dualistically corresponding singularities (point of inflexion, double point, inflexional tangent, double tangent) were extended by Cayley to
curves with higher singularities. By the aid of developments in series he derived four "equivalence numbers" which enable us to determine how many singularities are absorbed into a singular point of higher
order, and how the expression for the deficiency of
the curve is modified thereby. Cayley's results were
confirmed, extended, and completed as to proofs by
the works of Nöther, Zeuthen, Halphen, and Smith.
The fundamental question arising from the Cayley
method of considering the subject, whether and by
what change of parameters a curve with correspond-

ing elementary singularities can be derived from a
curve with higher singularity, for which the Plücker
and deficiency equations are the same, has been
studied by A. Brill.

Plücker's greatest service consisted in the introduction of the straight line as a space element.   The
principle of duality had led him to introduce, besides
the point in the plane, the straight line, and in space
the plane as a determining element.   Plücker also
used in space the straight line for the systematic generation of geometric figures.   His first works in this
direction were laid before the Royal Society in London in 1865.   They contained theorems on complexes,
congruences, and ruled surfaces with some indications
of the method of proof.   The further development
appeared in 1868 as *Neue Geometrie des Raumes, ge-
gründet auf die Betrachtung der geraden Linie als Raum-
element.*   Plücker had himself made a study of linear
complexes but his completion of the theory of complexes of the second degree was interrupted by death.
Further extension of the theory of complexes was
made by F. Klein.

The results contained in Plücker's last work have
thrown a flood of light upon the difference between
plane and solid geometry.   The curved line of the
plane appears as a simply infinite system either of
points or of straight lines; in space the curve may be
regarded as a simply infinite system of points, straight
lines or planes; but from another point of view this

curve in space may be replaced by the developable surface of which it is the edge of regression. Special cases of the curve in space and the developable surface are the plane curve and the cone. A further space figure, the general surface, is on the one side a doubly infinite system of points or planes, but on the other, as a special case of a complex, a triply infinite system of straight lines, the tangents to the surface. As a special case we have the skew surface or ruled surface. Besides this the congruence appears as a doubly, the complex as a triply, infinite system of straight lines. The geometry of space involves a number of theories to which plane geometry offers no analogy. Here belong the relations of a space curve to the surfaces which may be passed through it, or of a surface to the gauche curves lying upon it. To the lines of curvature upon a surface there is nothing corresponding in the plane, and in contrast to the consideration of the straight line as the shortest line between two points of a plane, there stand in space two comprehensive and difficult theories, that of the geodetic line upon a given surface and that of the minimal surface with a given boundary. The question of the analytic representation of a gauche curve involves peculiar difficulties, since such a figure can be represented by two equations between the co-ordinates x, y, z only when the curve is the complete intersection of two surfaces. In just this direction tend the modern investigations of Nöther, Halphen, and Valentiner.

Four years after the *Analytisch-geometrische Untersuchungen* of Plücker, in the year 1832, Steiner published his *Systematische Entwicklung der Abhängigkeit geometrischer Gestalten*. Steiner found the whole theory of conic sections concentrated in the single theorem (with its dualistic analogue) that a curve of the second order is produced as the intersection of two collinear or projective pencils, and hence the theory of curves and surfaces of the second order was essentially completed by him, so that attention could be turned to algebraic curves and surfaces of higher order. Steiner himself followed this course with good results. This is shown by the "Steiner surface," and by a paper which appeared in 1848 in the *Berliner Abhandlungen*. In this the theory of the polar of a point with respect to a curved line was treated exhaustively and thus a more geometric theory of plane curves developed, which was further extended by the labors of Grassmann, Chasles, Jonquières, and Cremona.*

The names of Steiner and Plücker are also united in connection with a problem which in its simplest form belongs to elementary geometry, but in its generalization passes into higher fields. It is the Malfatti Problem.† In 1803 Malfatti gave out the following problem: From a right triangular prism to cut out three cylinders which shall have the same altitude as the prism, whose volumes shall be the greatest possible, and consequently the mass remain-

*Loria.   † Wittstein, *Geschichte des Malfatti'schen Problems*, 1871.

ing after their removal shall be a minimum. This problem he reduced to what is now generally known as Malfatti's problem: In a
given triangle to inscribe three circles so that each circle shall be
tangent to two sides of the triangle and to the other two circles. He
calculates the radii x_1, x_2, x_3 of the circles sought in terms of the
semi-perimeter s of the triangle, the radius ρ of the inscribed circle, the distances a_1, a_2, a_3; b_1, b_2, b_3 of the vertices of the triangle from the center of the inscribed circle and its points of tangency to the sides, and gets:

$$x_1 = \frac{\rho}{2b_1}(s + a_1 - \rho - a_2 - a_3),$$

$$x_2 = \frac{\rho}{2b_2}(s + a_2 - \rho - a_3 - a_1),$$

$$x_3 = \frac{\rho}{2b_3}(s + a_3 - \rho - a_1 - a_2),$$

without giving the calculation in full; but he adds a simple construction. Steiner also studied this problem. He gave (without
proof) a construction, showed that there are thirty-two solutions
and generalized the problem, replacing the three straight lines by
three circles. Plücker also considered this same generalization.
But besides this Steiner studied the same problem for space: In
connection with three given conics upon a surface of the second
order to determine three others which shall each touch two of the
given conics and two of the required. This general problem received an analytic solution from Schellbach and Cayley, and also
from Clebsch with the aid of the addition theorem of elliptic functions, while the more simple problem in the plane was attacked in
the greatest variety of ways by Gergonne, Lehmus, Crelle, Grunert,
Scheffler, Schellbach (who gave a specially elegant trigonometric
solution) and Zorer. The first perfectly satisfactory proof of Steiner's construction was given by Binder.[*]

After Steiner came von Staudt and Chasles who
rendered excellent service in the development of pro-

[*] *Programm Schönthal, 1868.*

jective geometry. In 1837 Michel Chasles published
his *Aperçu historique sur l'origine et le développement
des méthodes en géométrie*, a work in which both ancient
and modern methods are employed in the derivation
of many interesting results, of which several of the
most important, among them the introduction of the
cross-ratio (Chasles's "anharmonic ratio") and the
reciprocal and collinear relation (Chasles's "duality"
and "homography"), are to be assigned in part to
Steiner and in part to Möbius.

Von Staudt's *Geometrie der Lage* appeared in 1847,
his *Beiträge zur Geometrie der Lage*, 1856–1860. These
works form a marked contrast to those of Steiner and
Chasles who deal continually with metric relations
and cross-ratios, while von Staudt seeks to solve the
problem of "making the geometry of position an independent science not standing in need of measurement." Starting from relations of position purely,
von Staudt develops all theorems that do not deal
immediately with the magnitude of geometric forms,
completely solving, for example, the problem of the
introduction of the imaginary into geometry. The
earlier works of Poncelet, Chasles, and others had,
to be sure, made use of complex elements but had
defined the same in a manner more or less vague and,
for example, had not separated conjugate complex
elements from each other. Von Staudt determined
the complex elements as double elements of involu-
tion-relations. Each double element is characterized

by the sense in which, by this relation, we pass from the one to the other. This suggestion of von Staudt's, however, did not become generally fruitful, and it was reserved for later works to make it more widely known by the extension of the originally narrow conception.

In the *Beiträge* von Staudt has also shown how the cross-ratios of any four elements of a prime form of the first class (von Staudt's *Würfe*) may be used to derive absolute numbers from pure geometry.*

With the projective geometry is most closely connected the modern descriptive geometry. The former in its development drew its first strength from the considerations of perspective, the latter enriches itself later with the fruits matured by the cultivation of projective geometry.

The perspective of the Renaissance† was developed especially by French mathematicians, first by Desargues who used co-ordinates in his pictorial representation of objects in such a way that two axes lay in the picture plane, while the third axis was normal to this plane. The results of Desargues were more important, however, for theory than for practice. More valuable results were secured by Taylor with a "linear perspective" (1715). In this a straight line is determined by its trace and vanishing point, a plane by its trace and vanishing line. This method was

* Stolz, O., *Vorlesungen über allgemeine Arithmetik*, 1885-1886.
† Wiener.

used by Lambert in an ingenious manner for different constructions, so that by the middle of the eighteenth century even space-forms in general position could be pictured in perspective.

Out of the perspective of the eighteenth century grew "descriptive geometry," first in a work of Frézier's, which besides practical methods contained a special theoretical section furnishing proofs for all cases of the graphic methods considered. Even in the "description," or representation, Frézier replaces the central projection by the perpendicular parallel-projection, "which may be illustrated by falling drops of ink."* The picture of the plane of projection is called the ground plane or elevation according as the picture plane is horizontal or vertical. With the aid of this "description" Frézier represents planes, polyhedra, surfaces of the second degree as well as intersections and developments.

Since the time of Monge descriptive geometry has taken rank as a distinct science. The *Leçons de géométrie descriptive* (1795) form the foundation-pillars of descriptive geometry, since they introduce horizontal and vertical planes with the ground-line and show how to represent points and straight lines by two projections, and planes by two traces. This is followed in the *Leçons* by the great number of problems of intersection, contact and penetration which arise from combinations of planes with polyhedra and surfaces

* Wiener.

of the second order. Monge's successors, Lacroix, Hachette, Olivier, and J. de la Gournerie applied these methods to surfaces of the second order, ruled surfaces, and the relations of curvature of curves and surfaces.

Just at this time, when the development of descriptive geometry in France had borne its first remarkable results, the technical high schools came into existence. In the year 1794 was established in Paris the *École Centrale des Travaux Publics* from which in 1795 the *École Polytechnique* was an outgrowth. Further technical schools, which in course of time attained to university rank, were founded in Prague in 1806, in Vienna in 1815, in Berlin in 1820, in Karlsruhe in 1825, in Munich in 1827, in Dresden in 1828, in Hanover in 1831, in Stuttgart 1832, in Zürich in 1860, in Braunschweig in 1862, in Darmstadt in 1869, and in Aix-la-Chapelle in 1870. In these institutions the results of projective geometry were used to the greatest advantage in the advancement of descriptive geometry, and were set forth in the most logical manner by Fiedler, whose text-books and manuals, in part original and in part translations from the English, take a conspicuous place in the literature of the science.

With the technical significance of descriptive geometry there has been closely related for some years an artistic side, and it is this especially which has marked an advance in works on axonometry (Weisbach, 1844), relief-perspective, photogrammetry, and theory of lighting.

The second quarter of our century marks the time when developments in form-theory in connection with geometric constructions have led to the discovery of of new and important results. Stimulated on the one side by Jacobi, on the other by Poncelet and Steiner,

Hesse (1837–1842) by an application of the transformation of homogeneous forms treated the theory of surfaces of the second order and constructed their principal axes.* By him the notions of "polar triangles" and "polar tetrahedra" and of "systems of conjugate points" were introduced as the geometric expression of analytic relations. To these were added the linear construction of the eighth intersection of three surfaces of the second degree, when seven of them are given, and also by the use of Steiner's theorems, the linear construction of a surface of the second degree from nine given points. Clebsch, following the English mathematicians, Sylvester, Cayley, and Salmon, went in his works essentially further than Hesse. His vast contributions to the theory of invariants, his introduction of the notion of the deficiency of a curve, his applications of the theory of elliptic and Abelian functions to geometry and to the study of rational and elliptic curves, secure for him a pre-eminent place among those who have advanced the science of extension. As an algebraic instrument Clebsch, like Hesse, had a fondness for the theorem upon the multiplication of determinants in its application to bordered determinants. His works† upon the general theory of algebraic curves and surfaces

---

*Nöther, "Otto Hesse," *Schlömilch's Zeitschrift*, Bd. 20, Hl. A.

†"Clebsch, Versuch einer Darlegung und Würdigung seiner wissenschaftlichen Leistungen von einigen seiner Freunde" (Brill, Gordan, Klein, Lüroth, A. Mayer, Nöther, Von der Mühll) *Math. Ann.*, Bd. 7.

began with the determination of those points upon an algebraic surface at which a straight line has four-point contact, a problem also treated by Salmon but not so thoroughly. While now the theory of surfaces of the third order with their systems of twenty-seven straight lines was making headway on English soil, Clebsch undertook to render the notion of "deficiency" fruitful for geometry. This notion, whose analytic properties were not unknown to Abel, is found first in Riemann's *Theorie der Abel'schen Funktionen* (1857). Clebsch speaks also of the deficiency of an algebraic curve of the nth order with d double points and r points of inflexion, and determines the number $p = \frac{1}{2}(n-1)(n-2) - d - r$. To one class of plane or gauche curves characterized by a definite value of p belong all those that can be made to pass over into one another by a rational transformation or which possess the property that any two have a one-to-one correspondence. Hence follows the theorem that only those curves that possess the same $3p - 3$ parameters (for curves of the third order, the same one parameter) can be rationally transformed into one another.

The difficult theory of gauche curves* owes its first general results to Cayley, who obtained formulae corresponding to Plücker's equations for plane curves. Works on gauche curves of the third and fourth orders had already been published by Möbius, Chasles, and Von Staudt. General observations on gauche curves

* Loria.

in more recent times are found in theorems of Nöther
and Halphen.

The foundations of enumerative geometry* are
found in Chasles's method of characteristics (1864).
Chasles determined for rational configurations of one
dimension a correspondence-formula which in the
simplest case may be stated as follows: If two ranges
of points R_1 and R_2 lie upon a straight line so that to
every point x of R_1 there correspond in general a
points y in R_2, and again to every point y of R_2 there
always correspond β points x in R_1, the configuration
formed from R_1 and R_2 has $(a+\beta)$ coincidences or
there are $(a+\beta)$ times in which a point x coincides
with a corresponding point y. The Chasles corre-
spondence-principle was extended inductively by Cayley in 1866 to point-systems of a curve of higher
deficiency and this extension was proved by Brill.†
Important extensions of these enumerative formulae
(correspondence-formulae), relating to general algebraic curves, have been given by Brill, Zeuthen, and
Hurwitz, and set forth in elegant form by the introduction of the notion of deficiency. An extended
treatment of the fundamental problem of enumerative
geometry, to determine how many geometric configurations of given definition satisfy a sufficient number
of conditions, is contained in the *Kalkül der abzählen-
den Geometrie* by H. Schubert (1879).

The simplest cases of one-to-one correspondence

---

* Loria.　　　　　　　　　† *Mathem. Annalen*, VI.

or uniform representation, are furnished by two planes superimposed one upon the other. These are the similarity studied by Poncelet and the collineation treated by Möbius, Magnus, and Chasles.[*] In both cases to a point corresponds a point, to a straight line a straight line. From these linear transformations Poncelet, Plücker, Magnus, Steiner passed to the quadratic where they first investigated one-to-one correspondences between two separate planes. The "Steiner projection" (1832) employed two planes E_1 and E_2 together with two straight lines g_1 and g_2 not co-planar. If we draw through a point P_1 or P_2 of E_1 or E_2 the straight line x_1 or x_2 which cuts g_1 as well as g_2, and determines the intersection X_2 or X_1, with E_2 or E_1, then are P_1 and X_2, and P_2 and X_1 corresponding points. In this manner to every straight line of the one plane corresponds a conic section in the other. In 1847 Plücker had determined a point upon the hyperboloid of one sheet, like fixing a point in the plane, by the segments cut off upon the two generators passing through the point by two fixed generators. This was an example of a uniform representation of a surface of the second order upon the plane.

The one-to-one relation of an arbitrary surface of the second order to the plane was investigated by Chasles in 1863, and this work marks the beginning of the proper theory of surface representation which

* Loria.

found its further development when Clebsch and Cremona independently succeeded in the representation of surfaces of the third order. Cremona's important results were extended by Cayley, Clebsch, Rosanes, and Nöther, to the last of whom we owe the important theorem that every Cremona transformation which as such is uniform forward and backward can be effected by the repetition of a number of quadratic transformations. In the plane only is the aggregate of all rational or Cremona transformations known; for the space of three dimensions, merely a beginning of the development of this theory has been made.*

A specially important case of one-to-one correspondence is that of a conformal representation of a surface upon the plane, because here similarity in the smallest parts exists between original and image. The simplest case, the stereographic projection, was known to Hipparchus and Ptolemy. The representation by reciprocal radii characterized by the fact that any two corresponding points P_1 and P_2 lie upon a ray through the fixed point O so that $OP_1 \cdot OP_2 =$ constant, is also conformal. Here every sphere in space is in general transformed into a sphere. This transformation, studied by Bellavitis 1836 and Stubbs 1843, is especially useful in dealing with questions of mathematical physics. Sir Wm. Thomson calls it "the principle of electric images." The investigations upon representa-

---

* Klein, F., *Vergleichende Betrachtungen über neuere geometrische Forschungen*, 1872.

tions, made by Lambert and Lagrange, but more especially those by Gauss, lead to the theory of curvature.

A further branch of geometry, the differential geometry (theory of curvature of surfaces), considers in general not first the surface in its totality but the properties of the same in the neighborhood of an ordinary point of the surface, and with the aid of the differential calculus seeks to characterize it by analytic formulae.

The first attempts to enter this domain were made by Lagrange (1761), Euler (1766), and Meusnier(1776). The former determined the differential equation of minimal surfaces ; the two latter discovered certain theorems upon radii of curvature and surfaces of centers. But of fundamental importance for this rich domain have been the investigations of Monge, Dupin, and especially of Gauss. In the *Application de l'analyse à la géométrie* (1795), Monge discusses families of surfaces (cylindrical surfaces, conical surfaces, and surfaces of revolution,—envelopes with the new notions of characteristic and edge of regression) and determines the partial differential equations distinguishing each. In the year 1813 appeared the *Développements de géométrie* by Dupin. It introduced the indicatrix at a point of a surface, as well as extensions of the theory of lines of curvature (introduced by Monge) and of asymptotic curves.

Gauss devoted to differential geometry three treatises: the most celebrated, *Disquisitiones generales circa superficies curvas*, appeared in 1827, the other two *Untersuchungen über Gegenstände der höheren Geodäsie* were published in 1843 and 1846. In the *Disquisitiones*, to the preparation of which he was led by his own astronomical and geodetic investigations,* the spherical representation of a surface is introduced. The one-to-one correspondence between the surface and the sphere is established by regarding as corresponding points the feet of parallel normals, where obviously we must restrict ourselves to a portion of the given surface, if the correspondence is to be maintained. Thence follows the introduction of the curvilinear co-ordinates of a surface, and the definition of the measure of curvature as the reciprocal of the product of the two radii of principal curvature at the point under consideration. The measure of curvature is first determined in ordinary rectangular co-ordinates and afterwards also in curvilinear co-ordinates of the surface. Of the latter expression it is shown that it is not changed by any bending of the surface without stretching or folding (that it is an invariant of curvature). Here belong the consideration of geodetic lines, the definition and a fundamental theorem upon the total curvature (*curvatura integra*) of a triangle bounded by geodetic lines.

The broad views set forth in the *Disquisitiones* of 1827 sent out fruitful suggestions in the most vari-

* Brill, A., *Antrittsrede in Tübingen*, 1884.

ous directions. Jacobi determined the geodetic lines of the general ellipsoid. With the aid of elliptic coordinates (the parameters of three surfaces of a system of confocal surfaces of the second order passing through the point to be determined) he succeeded in integrating the partial differential equation so that the equation of the geodetic line appeared as a relation between two Abelian integrals. The properties of the geodetic lines of the ellipsoid are derived with especial ease from the elegant formulae given by Liouville. By Lamé the theory of curvilinear co-ordinates, of which he had investigated a special case in 1837, was developed in 1859 into a theory for space in his *Leçons sur la théorie des coordonnées curvilignes.*

The expression for the Gaussian measure of curvature as a function of curvilinear co-ordinates has given an impetus to the study of the so-called differential invariants or differential parameters. These are certain functions of the partial derivatives of the coefficients in the expression for the square of the line-element which in the transformation of variables behave like the invariants of modern algebra. Here Saucé, Jacobi, C. Neumann, and Halphen laid the foundations, and a general theory has been developed by Beltrami.* This theory, as well as the contact-transformations of Lie, moves along the border line between geometry and the theory of differential equations.†

* *Mem. di Bologna,* VIII. † Loria.

With problems of the mathematical theory of light are connected certain investigations upon systems of rays and the properties of infinitely thin bundles of rays, as first carried on by Dupin, Malus, Ch. Sturm, Bertrand, Transon, and Hamilton. The
celebrated works of Kummer (1857 and 1866) perfect Hamilton's
results upon bundles of rays and consider the number of singularities of a system of rays and its focal surface. An interesting application to the investigation of the bundles of rays between the
lens and the retina, founded on the study of the infinitely thin
bundles of normals of the ellipsoid, was given by O. Böklen.*

Non-Euclidean Geometry.— Though the respect
which century after century had paid to the *Elements*
of Euclid was unbounded, yet mathematical acuteness
had discovered a vulnerable point; and this point†
forms the eleventh axiom (according to Hankel, reckoned by Euclid himself among the postulates) which
affirms that two straight lines intersect on that side of
a transversal on which the sum of the interior angles
is less than two right angles. Toward the end of the
last century Legendre had tried to do away with this
axiom by making its proof depend upon the others, but
his conclusions were invalid. This effort of Legendre's
was an indication of the search now beginning after a
geometry free from contradictions, a hyper-Euclidean
geometry or pangeometry. Here also Gauss was
among the first who recognized that this axiom could
not be proved. Although from his correspondence
with Wolfgang Bolyai and Schumacher it can easily

* *Kronecker's Journal*, Band 46. *Fortschritte*, 1884.
† Loria.

be seen that he had obtained some definite results in
this field at an early period, he was unable to decide
upon any further publication. The real pioneers in
the Non-Euclidean geometry were Lobachevski and
the two Bolyais. Reports of the investigations of
Lobachevski first appeared in the *Courier* of Kasan,
1829–1830, then in the transactions of the University of Kasan, 1835–1839, and finally as *Geometrische
Untersuchungen über die Theorie der Parallellinien*, 1840,
in Berlin. By Wolfgang Bolyai was published (1832–
1833*) a two-volume work, *Tentamen Juventutem stu-
diosam in elementa Matheseos purae, etc.* Both works
were for the mathematical world a long time as good
as non-existent till first Riemann, and then (in 1866)
R. Baltzer in his *Elemente*, referred to Bolyai. Almost
at the same time there followed a sudden mighty advance toward the exploration of this "new world" by
Riemann, Helmholtz, and Beltrami. It was recognized that of the twelve Euclidean axioms† nine are
of essentially arithmetic character and therefore hold
for every kind of geometry; also to every geometry is
applicable the tenth axiom upon the equality of all
right angles. The twelfth axiom (two straight lines,
or more generally two geodetic lines, include no
space) does not hold for geometry on the sphere.
The eleventh axiom (two straight lines, geodetic

* Schmidt, "Aus dem Leben zweier ungarischen Mathematiker," *Grunert
Arch.*, Bd. 48.

† Brill, A., *Ueber das elfte Axiom des Euclid*, 1883.

lines, intersect when the sum of the interior angles is less than two right angles) does not hold for geometry on a pseudo-sphere, but only for that in the plane.

Riemann, in his paper "Ueber die Hypothesen, welche der Geometrie zu Grunde liegen,"[*] seeks to penetrate the subject by forming the notion of a multiply extended manifoldness; and according to these investigations the essential characteristics of an n-ply extended manifoldness of constant measure of curvature are the following:

1. "Every point in it may be determined by n variable magnitudes (co-ordinates).

2. "The length of a line is independent of position and direction, so that every line is measurable by every other.

3. "To investigate the measure-relations in such a manifoldness, we must for every point represent the line-elements proceeding from it by the corresponding differentials of the co-ordinates. This is done by virtue of the hypothesis that the length-element of the line is equal to the square root of a homogeneous function of the second degree of the differentials of the co-ordinates."

At the same time Helmholtz[†] published in the "Thatsachen, welche der Geometrie zu Grunde lie gen," the following postulates:

---

*Göttinger Abhandlungen, XIII., 1868. Fortschritte, 1868.

†Fortschritte, 1868.

1. "A point of an n-tuple manifoldness is determined by n co-ordinates.

2. "Between the $2n$ co-ordinates of a point-pair there exists an equation, independent of the movement of the latter, which is the same for all congruent point-pairs.

3. "Perfect mobility of rigid bodies is assumed.

4. "If a rigid body of n dimensions revolves about $n-1$ fixed points, then revolution without reversal will bring it back to its original position."

Here spatial geometry has satisfactory foundations for a development free from contradictions, if it is further assumed that space has three dimensions and is of unlimited extent.

One of the most surprising results of modern geometric investigations was the proof of the applicability of the non-Euclidean geometry to pseudo-spheres or surfaces of constant negative curvature.* On a pseudo-sphere, for example, it is true that a geodetic line (corresponding to the straight line in the plane, the great circle on the sphere) has two separate points at infinity; that through a point P, to a given geodetic line g, there are two parallel geodetic lines, of which, however, only one branch beginning at P cuts g at infinity while the other branch does not meet g at all; that the sum of the angles of a geodetic triangle is less than two right angles. Thus we have a geometry upon the pseudo-sphere which with the spherical ge-

---

* Cayley, *Address to the British Association*, etc., 1883.

ometry has a common limiting case in the ordinary
or Euclidean geometry. These three geometries have
this in common that they hold for surfaces of constant
curvature. According as the constant value of the
curvature is positive, zero, or negative, we have to do
with spherical, Euclidean, or pseudo-spherical geometry.

A new presentation of the same theory is due to
F. Klein. After projective geometry had shown that
in projection or linear transformation all descriptive
properties and also some metric relations of the figures remain unaltered, the endeavor was made to find
for the metric properties an expression which should
remain invariant after a linear transformation. After
a preparatory work of Laguerre which made the "notion of the angle projective," Cayley, in 1859, found the
general solution of this problem by considering "every
metric property of a plane figure as contained in a
projective relation between it and a fixed conic."
Starting from the Cayley theory, on the basis of the
consideration of measurements in space, Klein succeeded in showing that from the projective geometry
with special determination of measurements in the
plane there could be derived an elliptic, parabolic,
or hyperbolic geometry,* the same fundamentally as
the spherical, Euclidean, or pseudo-spherical geometry respectively.

The need of the greatest possible generalization

* *Fortschritte*, 1871.

and the continued perfection of the analytic apparatus have led to the attempt to build up a geometry of n dimensions; in this, however, only individual relations have been considered. Lagrange* observes that "mechanics may be regarded as a geometry of four dimensions." Plücker endeavored to clothe the notion of arbitrarily extended space in a form easily understood. He showed that for the point, the straight line or the sphere, the surface of the second order, as a space element, the space chosen must have three, four, or nine dimensions respectively. The first investigation, giving a different conception from Plücker's and "considering the element of the arbitrarily extended manifoldness as an analogue of the point of space," is found† in H. Grassmann's principal work, *Die Wissenschaft der extensiven Grösse oder die lineale Ausdehnungslehre* (1844), which remained almost wholly unnoticed, as did his *Geometrische Analyse* (1847). Then followed Riemann's studies in multiply extended manifoldnesses in his paper *Ueber die Hypothesen*, etc., and they again furnished the starting point for a series of modern works by Veronese, H. Schubert, F. Meyer, Segre, Castelnuovo, etc.

A *Geometria situs* in the broader sense was created by Gauss, at least in name; but of it we know scarcely more than certain experimental truths.‡ The *Analysis*

---

* Loria.

† F. Klein, *Vergleichende Betrachtungen über neuere geometrische Forschungen*, 1872.

‡ Brill, A., *Antrittsrede in Tübingen*, 1884.

situs, suggested by Riemann, seeks what remains fixed after transformations consisting of the combination of infinitesimal distortions.* This aids in the solution of problems in the theory of functions. The contact transformations already considered by Jacobi have been developed by Lie. A contact-transformation is defined analytically by every substitution which expresses the values of the co-ordinates x, y, z, and the partial derivatives $\frac{dz}{dx} = p$, $\frac{dz}{dy} = q$, in terms of quantities of the same kind, x', y', z', p', q'. In such a transformation contacts of two figures are replaced by similar contacts.

Also a "geometric theory of probability" has been created by Sylvester and Woolhouse;† Crofton uses it for the theory of lines drawn at random in space.

In a history of elementary mathematics there possibly calls for attention a related field, which certainly cannot be regarded as a branch of science, but yet which to a certain extent reflects the development of geometric science, the history of geometric illustrative material.‡ Good diagrams or models of systems of space-elements assist in teaching and have frequently led to the rapid spread of new ideas. In fact in the geometric works of Euler, Newton, and Cramer are found numerous plates of figures. Interest in the

---

*F. Klein.   † *Fortschritte*, 1868.

‡ Brill, A., *Ueber die Modellsammlung des mathematischen Seminars der Universität Tübingen*, 1886. *Mathematisch-naturwissenschaftliche Mittheilungen* von O. Böklen. 1887.

construction of models seems to have been manifested first in France in consequence of the example and activity of Monge. In the year 1830 the *Conservatoire des arts et métiers* in Paris possessed a whole series of thread models of surfaces of the second degree, conoids and screw surfaces. A further advance was made by Bardin (1855). He had plaster and thread models constructed for the explanation of stone-cutting, toothed gears and other matters. His collection was considerably enlarged by Muret. These works of French technologists met with little acceptance from the mathematicians of that country, but, on the contrary, in England Cayley and Henrici put on exhibition in London in 1876 independently constructed models together with other scientific apparatus of the universities of London and Cambridge.

In Germany the construction of models experienced an advance from the time when the methods of projective geometry were introduced into descriptive geometry. Plücker, who in his drawings of curves of the third order had in 1835 showed his interest in relations of form, brought together in 1868 the first large collection of models. This consisted of models of complex surfaces of the fourth order and was considerably enlarged by Klein in the same field. A special surface of the fourth order, the wave-surface for optical bi-axial crystals was constructed in 1840 by Magnus in Berlin, and by Soleil in Paris. In the year 1868 appeared the first model of a surface of the third order with its twenty-seven straight lines, by Chr. Wiener. In the sixties, Kummer constructed models of surfaces of the fourth order and of certain focal surfaces. His pupil Schwarz likewise constructed a series of models, among them minimal surfaces and the surfaces of centers of the ellipsoid. At a meeting of mathematicians in Göttingen there was made a notable exhibition of models which stimulated further work in this direction.

In wider circles the works suggested by A. Brill, F. Klein, and W. Dyck in the mathematical seminar of the Munich polytechnic school have found recognition. There appeared from 1877 to 1890 over a hundred models of the most various kinds, of value not only in mathematical teaching but also in lectures on perspective, mechanics and mathematical physics.

In other directions also has illustrative material of this sort been multiplied, such as surfaces of the third order by Rodenberg, thread models of surfaces and gauche curves of the fourth order by Rohn, H. Wiener, and others.

*   *   *

If one considers geometric science as a whole, it cannot be denied that in its field no essential difference between modern analytic and modern synthetic geometry any longer exists. The subject matter and the methods of proof in both directions have gradually taken almost the same form. Not only does the synthetic method make use of space intuition; the analytic representations also are nothing less than a clear expression of space relations. And since metric properties of figures may be regarded as relations of the same to a fundamental form of the second order, to the great circle at infinity, and thus can be brought into the aggregate of projective properties, instead of analytic and synthetic geometry, we have only a projective geometry which takes the first place in the science of space.[*]

The last decades, especially of the development of German mathematics, have secured for the science a leading place. In general two groups of allied works may be recognized.[†] In the treatises of the one tendency "after the fashion of a Gauss or a Dirichlet, the inquiry is concentrated upon the exactest possible limitation of the fundamental notions" in the theory of functions, theory of numbers, and mathematical physics. The investigations of the other tendency, as is to be seen in Jacobi and Clebsch, start "from a small circle of already recognized fundamental concepts and aim at the relations and consequences which spring from them," so as to serve modern algebra and geometry.

On the whole, then, we may say that[‡] "mathematics have steadily advanced from the time of the Greek geometers. The achievements of Euclid, Archimedes, and Apollonius are as admirable now as they

---

* F. Klein.　　　　　† Clebsch.
‡ Cayley, A., *Address to the British Association*, etc., 1883.

were in their own days. Descartes's method of co-
ordinates is a possession forever. But mathematics
have never been cultivated more zealously and dili-
gently, or with greater success, than in this century—
in the last half of it, or at the present time: the advances made have been enormous, the actual field is
boundless, the future is full of hope."

V. TRIGONOMETRY.

A. GENERAL SURVEY.

TRIGONOMETRY was developed by the ancients for purposes of astronomy. In the first period a number of fundamental formulae of trigonometry were established, though not in modern form, by the Greeks and Arabs, and employed in calculations. The second period, which extends from the time of the gradual rise of mathematical sciences in the earliest Middle Ages to the middle of the seventeenth century, establishes the science of calculation with angular functions and produces tables in which the sexagesimal division is replaced by decimal fractions, which marks a great advance for the purely numerical calculation. During the third period, plane and spherical trigonometry develop, especially polygonometry and polyhedrometry which are almost wholly new additions to the general whole. Further additions are the projective formulae which have furnished a series of interesting results in the closest relation to projective geometry.

B. FIRST PERIOD.

The Papyrus of Ahmes* speaks of a quotient called *seqt*. After observing that the great pyramids all possess approximately equal angles of inclination, the assumption is rendered probable that this *seqt* is identical with the cosine of the angle which the edge of the pyramid forms with the diagonal of the square base. This angle is usually 52°. In the Egyptian monuments which have steeper sides, the *seqt* appears to be equal to the trigonometric tangent of the angle of inclination of one of the faces to the base.

Trigonometric investigations proper appear first among the Greeks. Hypsicles gives the division of the circumference into three hundred sixty degrees, which, indeed, is of Babylonian origin but was first turned to advantage by the Greeks. After the introduction of this division of the circle, sexagesimal fractions were to be found in all the astronomical calculations of antiquity (with the single exception of Heron), till finally Peurbach and Regiomontanus prepared the way for the decimal reckoning. Hipparchus was the first to complete a table of chords, but of this we have left only the knowledge of its former exist-

* Cantor, I., p. 58.

ence. In Heron are found actual trigonometric formulae with numerical ratios for the calculation of the
areas of regular polygons and in fact all the values of
$\cot\left(\dfrac{2\pi}{n}\right)$ for $n = 3, 4, \ldots 11, 12$ are actually computed.*
Menelaus wrote six books on the calculation of chords,
but these, like the tables of Hipparchus, are lost. On
the contrary, three books of the *Spherics* of Menelaus
are known in Arabic and Hebrew translations. These
contain theorems on transversals and on the congruence of spherical as well as plane triangles, and for
the spherical triangle the theorem that $a + b + c < 4R$,
$a + \beta + \gamma > 2R$.

The most important work of Ptolemy consists in
the introduction of a formal spherical trigonometry
for astronomical purposes. The thirteen books of the
Great Collection which contain the Ptolemaic astronomy and trigonometry were translated into Arabic,
then into Latin, and in the latter by a blending of the
Arabic article *al* with a Greek word arose the word
Almagest, now generally applied to the great work of
Ptolemy. Like Hypsicles, Ptolemy also, after the
ancient Babylonian fashion, divides the circumference into three hundred sixty degrees, but he, in addition to this, bisects every degree. As something
new we find in Ptolemy the division of the diameter
of the circle into one hundred twenty equal parts,
from which were formed after the sexagesimal fashion

* Tannery in *Mém. Bord.*, 1881.

two classes of subdivisions. In the later Latin translations these sixtieths of the first and second kind
were called respectively *partes minutae primae* and
partes minutae secundae. Hence came the later terms
"minutes" and "seconds." Starting from his theorem upon the inscribed quadrilateral, Ptolemy calculates the chords of arcs at intervals of half a degree.
But he develops also some theorems of plane and
especially of spherical trigonometry, as for example
theorems regarding the right angled spherical triangle.

A further not unimportant advance in trigonometry is to be noted in the works of the Hindus. The
division of the circumference is the same as that of
the Babylonians and Greeks; but beyond that there
is an essential deviation. The radius is not divided
sexagesimally after the Greek fashion, but the arc of
the same length as the radius is expressed in minutes; thus for the Hindus $r = 3438$ minutes. Instead
of the whole chords (*jiva*), the half chords (*ardhajya*)
are put into relation with the arc. In this relation of
the half-chord to the arc we must recognize the most
important advance of trigonometry among the Hindus.
In accordance with this notion they were therefore
familiar with what we now call the sine of an angle.
Besides this they calculated the ratios corresponding
to the versed sine and the cosine and gave them special names, calling the versed sine *utkramajya*, the
cosine *kotijya*. They also knew the formula $\sin^2 a$

$+ \cos^2 a = 1$. They did not, however, apply their trigonometric knowledge to the solution of plane triangles, but with them trigonometry was inseparably connected with astronomical calculations.

As in the rest of mathematical science, so in trigonometry, were the Arabs pupils of the Hindus, and still more of the Greeks, but not without important devices of their own. To Al Battani it was well known that the introduction of half chords instead of whole chords, as these latter appear in the *Almagest*, and therefore reckoning with the sine of an angle, is of essential advantage in the applications. In addition to the formulae found in the *Almagest*, Al Battani gives the relation, true for the spherical triangle, $\cos a = \cos b \cos c + \sin b \sin c \cos a$. In the consideration of right-angled triangles in connection with shadow-measuring, we find the quotients $\frac{\sin a}{\cos a}$ and $\frac{\cos a}{\sin a}$. These were reckoned for each degreee by Al Battani and arranged in a small table. Here we find the beginnings of calculation with tangents and cotangents. These names, however, were introduced much later. The origin of the term "sine" is due to Al Battani. His work upon the motion of the stars[*] was translated into Latin by Plato of Tivoli, and this translation contains the word *sinus* for half chord. In Hindu the half chord was called *ardhajya* or also *jiva* (which was used originally only for the whole

---

* Cantor, I., p. 693, where this account is considered somewhat doubtful.

chord); the latter word the Arabs adopted, simply
by reason of its sound, as *jiba*. The consonants of
this word, which in Arabic has no meaning of its
own, might be read *jaib* = bosom, or incision, and
this pronunciation, which apparently was naturalized
comparatively soon by the Arabs, Plato of Tivoli
translated properly enough into *sinus*. Thus was introduced the first of the modern names of the trigonometric functions.

Of astronomical tables there was no lack at that
time. Abul Wafa, by whom the ratio $\frac{\sin a}{\cos a}$ was called
the "shadow" belonging to the angle a, calculated a
table of sines at intervals of half a degree and also a
table of tangents, which however was used only for
determining the altitude of the sun.   About the same
time appeared the hakimitic table of sines which Ibn
Yunus of Cairo was required to construct by direction
of the Egyptian ruler Al Hakim.*

Among the Western Arabs the celebrated astronomer Jabir ibn Aflah, or Geber, wrote a complete trigonometry (principally spherical) after a method of his
own, and this work, rigorous throughout in its proofs,
was published in the Latin edition of his *Astronomy*
by Gerhard of Cremona.   This work contains a collection of formulae upon the right-angled spherical
triangle.   In the plane trigonometry he does not go

---

*Cantor, I., p. 743.

beyond the Almagest, and hence he here deals only with whole chords, just as Ptolemy had taught.

C. SECOND PERIOD.

FROM THE MIDDLE AGES TO THE MIDDLE OF THE SEVEN-TEENTH CENTURY.

Of the mathematicians outside of Germany in this period, Vieta made a most important advance by his introduction of the reciprocal triangle of a spherical triangle. In Germany the science was advanced by Regiomontanus and in its elements was presented with such skill and thorough knowledge that the plan laid out by him has remained in great part up to the present day. Peurbach had already formed the plan of writing a trigonometry but was prevented by death. Regiomontanus was able to carry out Peurbach's idea by writing a complete plane and spherical trigonometry. After a brief geometric introduction Regiomontanus's trigonometry begins with the right-angled triangle, the formulae needed for its computation being derived in terms of the sine alone and illustrated by numerical examples. The theorems on the right-angled triangle are used for the calculation of the equilateral and isosceles triangles. Then follow the principal cases of the oblique-angled triangle of which the first (a from a, b, c) is treated with much detail. The second book contains the sine theorem and a series of problems relating to triangles. The third, fourth, and fifth books bring in spherical trigonometry with many resemblances to Menelaus; in particular the angles are found from the sides. The case of the plane triangle (a from a, b, c), treated with considerable prolixity by Regiomontanus, received a shorter treatment from Rhaeticus, who established the formula $\cot \frac{1}{2} a = \dfrac{s - a}{\rho}$, where ρ is the radius of the inscribed circle.

In this period were also published Napier's equations, or analogies. They express a relation between the sum or difference of two sides (angles) and the third side (angle) and the sum or difference of the two opposite angles (sides).

Of modern terms, as already stated, the word "sine" is the oldest. About the end of the sixteenth century, or the beginning of the seventeenth, the abbreviation cosine for *complementi sinus* was introduced by the Englishman Gunter (died 1626). The terms tangent and secant were first used by Thomas Finck (1583); the term versed sine was used still earlier.*

By some writers of the sixteenth century, e. g., by Apian, *sinus rectus secundus* was written instead of cosine. Rhaeticus and Vieta have *perpendiculum* and *basis* for sine and cosine.† The natural values of the cosine, whose logarithms were called by Kepler "anti-

*Baltzer, R., *Die Elemente der Mathematik*, 1885.

†Pfleiderer, *Trigonometrie*, 1802.

logarithms," are first found calculated in the trigonometry of Copernicus as published by Rhaeticus.*

The increasing skill in practical computation, and the need of more accurate values for astronomical purposes, led in the sixteenth century to a strife after the most complete trigonometric tables possible. The preparation of these tables, inasmuch as the calculations were made without logarithms, was very tedious. Rhaeticus alone had to employ for this purpose a number of computers for twelve years and spent thereby thousands of gulden.†

The first table of sines of German origin is due to Peurbach. He put the radius equal to 600 000 and computed at intervals of 10' (in Ptolemy $r = 60$, with some of the Arabs $r = 150$). Regiomontanus computed two new tables of sines, one for $r = 6\,000\,000$, the other, of which no remains are left, for $r = 10\,000\,000$. Besides these we have from Regiomontanus a table of tangents for every degree, $r = 100\,000$. The last two tables evidently show a transition from the sexagesimal system to the decimal. A table of sines for every minute, with $r = 100\,000$, was prepared by Apian.

In this field should also be mentioned the indefatigable perseverance of Joachim Rhaeticus. He did not associate the trigonometric functions with the arcs of circles, but started with the right-angled tri-

* M. Curtze, in *Schlömilch's Zeitschrift*, Bd. XX.
† Gerhardt, *Geschichte der Mathematik in Deutschland* 1877.

angle and used the terms *perpendiculum* for sine, *basis*
for cosine.   He calculated (partly himself and partly
by the help of others) the first table of secants; later,
tables of sines, tangents, and secants for every 10″,
for radius $= 10\,000$ millions, and later still, for $r =$
10^{15}.   After his death the whole work was published
by Valentin Otho in the year 1596 in a volume of 1468
pages.*

To the calculation of natural trigonometric functions Bartholomaeus Pitiscus also devoted himself.
In the second book of his *Trigonometry* he sets forth
his views on computations of this kind.   His tables
contain values of the sines, tangents, and secants on
the left, and of the complements of the sines, tangents
and secants (for so he designated the cosines, cotangents, and cosecants) on the right.   There are added
proportional parts for 1′, and even for 10″.   In the
whole calculation r is assumed equal to 10^{25}.   The
work of Pitiscus appeared at the beginning of the
seventeenth century.

The tables of the numerical values of the trigonometric functions had now attained a high degree of
accuracy, but their real significance and usefulness
were first shown by the introduction of logarithms.

Napier is usually regarded as the inventor of logarithms, although Cantor's review of the evidence†
leaves no room for doubt that Bürgi was an independent discoverer.   His *Progress Tabulen*, computed be-

* Gerhardt.   † Cantor, II., pp. 662 et seq.

tween 1603 and 1611 but not published until 1620 is really a table of antilogarithms. Bürgi's more general point of view should also be mentioned. He desired to simplify all calculations by means of logarithms while Napier used only the logarithms of the trigonometric functions.

Bürgi was led to this method of procedure by comparison of the two series 0, 1, 2, 3, . . . and 1, 2, 4, 8, . . . or 2^0, 2^1, 2^2, 2^3, . . . He observed that for purposes of calculation it was most convenient to select 10 as the base of the second series, and from this standpoint he computed the logarithms of ordinary numbers, though he first decided on publication when Napier's renown began to spread in Germany by reason of Kepler's favorable reports. Bürgi's *Geometrische Progress Tabulen* appeared at Prague in 1620,* and contained the logarithms of numbers from 10^8 to 10^9 by tens. Bürgi did not use the term *logarithmus*, but by reason of the way in which they were printed he called the logarithms "red numbers," the numbers corresponding, "black numbers."

Napier started with the observation that if in a circle with two perpendicular radii OA_0 and OA_1 ($r = 1$), the sine $S_0S_1 \parallel OA_1$ moves from O to A_0 at intervals forming an arithmetic progression, its value decreases in geometrical progression. The segment OS_0, Napier originally called *numerus artificialis* and later the direction number or *logarithmus*. The first

* Gerhardt.

publication of this new method of calculation, in which $r = 10^7$, log sin $60^0 = 0$, log sin $0^0 = \infty$, so that the log-arithms increased as the sines decreased, appeared in 1614 and produced a great sensation. Henry Briggs had studied Napier's work thoroughly and made the important observation that it would be more suitable for computation if the logarithms were allowed to in-crease with the numbers. He proposed to put log 1 $= 0$, log 10 $= 1$, and Napier gave his assent. The ta-bles of logarithms calculated by Briggs, on the basis of this proposed change, for the natural numbers from 1 to 20 000 and from 90 000 to 100 000 were reckoned to 14 decimal places. The remaining gap was filled by the Dutch bookseller Adrian Vlacq. His tables which appeared in the year 1628 contained the logarithms of numbers from 1 to 100 000 to 10 decimal places. In these tables, under the name of his friend De Decker, Vlacq introduced logarithms upon the continent. As-sisted by Vlacq and Gellibrand, Briggs computed a table of sines to fourteen places and a table of tan-gents and secants to ten places, at intervals of 36″. These tables appeared in 1633. Towards the close of the seventeenth century Claas Vooght published a table of sines, tangents, and secants with their loga-rithms, and, what was especially remarkable, they were engraved on copper.

Thus was produced a collection of tables for logarithmic com-putation valuable for all time. This was extended by the intro-duction of the addition and subtraction logarithms always named after Gauss, but whose inventor, according to Gauss's own testimony, is Leonelli. The latter had proposed calculating a table with fourteen decimals; Gauss thought this impracticable, and calculated for his own use a table with five decimals.[*]

In the year 1875 there were in existence 553 logarithmic tables with decimal places ranging in number from 3 to 102. Arranged according to frequency, the 7-place tables stand at the head, then follow those with 5-places, 6-places, 4-places, and 10-places. The only table with 102 places is found in a work by H. M. Parkhurst (*Astronomical Tables*, New York, 1871).

Investigations of the errors occurring in logarithmic tables have been made by J. W. L. Glaisher.[†] It was there shown that every complete table had been transcribed, directly or indirectly after a more or less careful revision, from the table published in 1628 which contains the results of Briggs's *Arithmetica logarithmica* of 1624 for numbers from 1 to 100000 to ten places. In the first seven places Glaisher found 171 errors of which 48 occur in the interval from 1 to 10000. These errors, due to Vlacq, have gradually disappeared. Of the mistakes in Vlacq, 98 still appear in Newton (1658), 19 in Gardiner (1742), 5 in Vega (1797), 2 in Callet (1855), 2 in Sang (1871). Of the tables tested by Glaisher, four turned out to be free from error, viz., those of Bremiker (1857), Schrön (1860), Callet (1862), and Bruhns (1870). Contributions to the rapid calculation of common logarithms have been made by Koralek (1851) and R. Hoppe (1876); the work of the latter is based upon the theorem that every positive number may be transformed into an infinite product.[‡]

[*] Gauss, *Werke*, III., p. 244. Porro in *Bonc. Bull.*, XVIII.

[†] *Fortschritte*, 1873.

[‡] Stolz, *Vorlesungen über allgemeine Arithmetik*, 1885–1886.

D. THIRD PERIOD.

After Regiomontanus had laid the foundations of
plane and spherical trigonometry, and his successors
had made easier the work of computation by the computation of the numerical values of the trigonometric functions and the creation of a serviceable system of logarithms, the inner structure of the science
was ready to be improved in details during this third
period.   Important innovations are especially due to
Euler, who derived the whole of spherical trigonometry from a few simple theorems.   Euler defined the
trigonometric functions as mere numbers, so as to be
able to substitute them for series in whose terms appear arcs of circles from which the trigonometric functions proceed according to definite laws.   From him
we have a number of trigonometric formulae, in part
entirely new, and in part perfected in expression.
These were made especially clear when Euler denoted
the elements of the triangle by a, b, c, α, β, γ.   Then
such expressions as $\sin a$, $\tan a$ could be introduced
where formerly special letters had been used for the
same purpose.* Lagrange and Gauss restricted themselves to a single theorem in the derivation of spherical trigonometry.   The system of equations

---

* Baltzer, R., *Die Elemente der Mathematik*, 1885.

$$\sin\frac{a}{2}\,\sin\frac{b+c}{2}=\sin\frac{a}{2}\,\cos\frac{\beta-\gamma}{2},$$

with the corresponding relations, is ordinarily ascribed to Gauss, though the equations were first published by Delambre in 1807 (by Mollweide 1808, by Gauss 1809).* The case of the Pothenot problem is similar: it was discussed by Snellius 1614, by Pothenot 1692, by Lambert 1765.†

The principal theorems of polygonometry and polyhedrometry were established in the eighteenth century. To Euler we owe the theorem on the area of the orthogonal projection of a plane figure upon another plane; to Lexell the theorem upon the projection of a polygonal line. Lagrange, Legendre, Carnot and others stated trigonometric theorems for polyhedra (especially the tetrahedra), Gauss for the spherical quadrilateral.

The nineteenth century has given to trigonometry a series of new formulae, the so-called projective formulae. Besides Poncelet, Steiner, and Gudermann, Möbius deserves special mention for having devised a generalization of spherical trigonometry, such that sides or angles of a triangle may exceed 180°. The important improvements which in modern times trigonometric developments have contributed to other mathematical sciences, may be indicated in this one sentence: their extended description would considerably encroach upon the province of other branches of science.

* Hammer, *Lehrbuch der ebenen und sphärischen Trigonometrie*, 1897.
† Baltzer, R., *Die Elemente der Mathematik*, 1885.

BIOGRAPHICAL NOTES.*

Abel, Niels Henrik. Born at Findöe, Norway, August 5, 1802; died April 6, 1829. Studied in Christiania, and for a short time in Berlin and Paris. Proved the impossibility of the algebraic solution of the quintic equation; elaborated the theory of elliptic functions; founded the theory of Abelian functions.

Abul Jud, Mohammed ibn al Lait al Shanni. Lived about 1050. Devoted much attention to geometric problems not soluble with compasses and straight edge alone.

Abul Wafa al Buzjani. Born at Buzjan, Persia, June 10, 940; died at Bagdad, July 1, 998. Arab astronomer. Translated works of several Greek mathematicians; improved trigonometry and computed some tables; interested in geometric constructions requiring a single opening of the compasses.

Adelard. About 1120. English monk who journeyed through Asia Minor, Spain, Egypt, and Arabia. Made the first translation of Euclid from Arabic into Latin. Translated part of Al Khowarazmi's works.

Al Battani (Albategnius). Mohammed ibn Jabir ibn Sinan Abu Abdallah al Battani. Born in Battan, Mesopotamia, *c.* 850; died in Damascus, 929. Arab prince, governor of Syria; great-

---

*The translators feel that these notes will be of greater value to the reader by being arranged alphabetically than, as in the original, by periods, especially as this latter arrangement is already given in the body of the work. They also feel that they will make the book more serviceable by changing the notes as set forth in the original, occasionally eliminating matter of little consequence, and frequently adding to the meagre information given. They have, for this purpose, freely used such standard works as Cantor, Hankel, Günther, Zeuthen, et al., and especially the valuable little *Zeittafeln zur Geschichte der Mathematik, Physik und Astronomie bis zum Jahre 1500*, by Felix Müller, Leipzig, 1892. Dates are A. D., except when prefixed by the negative sign.

est Arab astronomer and mathematician. Improved trigonometry and computed the first table of cotangents.

Alberti, Leo Battista. 1404–1472. Architect, painter, sculptor.

Albertus Magnus. Count Albrecht von Bollstädt. Born at Lau-
ingen in Bavaria, 1193 or 1205; died at Cologne, Nov. 15,
1280. Celebrated theologian, chemist, physicist, and mathematician.

Al Biruni, Abul Rihan Mohammed ibn Ahmed. From Birun,
valley of the Indus; died 1038. Arab, but lived and travelled
in India and wrote on Hindu mathematics. Promoted spherical trigonometry.

Alcuin. Born at York, 736; died at Hersfeld, Hesse, May 19.
804. At first a teacher in the cloister school at York; then
assisted Charlemagne in his efforts to establish schools in
France.

Alhazen, Ibn al Haitam. Born at Bassora, 950; died at Cairo
1038. The most important Arab writer on optics.

Al Kalsadi, Abul Hasan Ali ibn Mohammed. Died 1486 or 1477.
From Andalusia or Granada. Arithmetician.

Al Karkhi, Abu Bekr Mohammed ibn al Hosain. Lived about
1010. Arab mathematician at Bagdad. Wrote on arithmetic,
algebra and geometry.

Al Khojandi, Abu Mohammed. From Khojand, in Khorassan;
was living in 992. Arab astronomer.

Al Khowarazmi, Abu Jafar Mohammed ibn Musa. First part of
ninth century. Native of Khwarazm (Khiva). Arab mathematician and astronomer. The title of his work gave the name
to algebra. Translated certain Greek works.

Al Kindi, Jacob ibn Ishak, Abu Yusuf. Born *c.* 813; died 873.
Arab philosopher, physician, astronomer and astrologer.

Al Kuhi, Vaijan ibn Rustam Abu Sahl. Lived about 975. Arab
astronomer and geometrician at Bagdad.

Al Nasawi, Abul Hasan Ali ibn Ahmed. Lived about 1000.
From Nasa in Khorassan. Arithmetician.

Al Sagani, Ahmed ibn Mohammed al Sagani Abu Hamid al Us-
turlabi. From Sagan, Khorassan; died 990. Bagdad astronomer

Anaxagoras. Born at Clazomene, Ionia, —499; died at Lampsacus, —428. Last and most famous philosopher of the Ionian school. Taught at Athens. Teacher of Euripides and Pericles.

Apianus (Apian), Petrus. Born at Leisnig, Saxony, 1495; died in 1552. Wrote on arithmetic and trigonometry.

Apollonius of Perga, in Pamphylia. Taught at Alexandria between —250 and —200, in the reign of Ptolemy Philopator. His eight books on conics gave him the name of "the great geometer." Wrote numerous other works. Solved the general quadratic with the help of conics.

Arbogast, Louis François Antoine. Born at Mutzig, 1759; died 1803. Writer on calculus of derivations, series, gamma function, differential equations.

Archimedes. Born at Syracuse, —287(?); killed there by Roman soldiers in —212. Engineer, architect, geometer, physicist. Spent some time in Spain and Egypt. Friend of King Hiero. Greatly developed the knowledge of mensuration of geometric solids and of certain curvilinear areas. In physics he is known for his work in center of gravity, levers, pulley and screw, specific gravity, etc.

Archytas. Born at Tarentum —430; died —365. Friend of Plato, a Pythagorean, a statesman and a general. Wrote on proportion, rational and irrational numbers, tore surfaces and sections, and mechanics.

Argand, Jean Robert. Born at Geneva, 1768; died *c.* 1825. Private life unknown. One of the inventors of the present method of geometrically representing complex numbers (1806).

Aristotle. Born at Stageira, Macedonia, —384; died at Chalcis, Euboea, —322. Founder of the peripatetic school of philosophy; teacher of Alexander the Great. Represented unknown quantities by letters; distinguished between geometry and geodesy; wrote on physics; suggested the theory of combinations.

Aryabhatta. Born at Pataliputra on the Upper Ganges, 476. Hindu mathematician. Wrote chiefly on algebra, including quadratic equations, permutations, indeterminate equations, and magic squares.

August, Ernst Ferdinand. Born at Prenzlau, 1795; died 1870 as director of the Kölnisch Realgymnasium in Berlin.

Autolykus of Pitane, Asia Minor. Lived about —330. Greek astronomer; author of the oldest work on spherics.

Avicenna, Abu Ali Hosain ibn Sina. Born at Charmatin, near Bokhara, 978; died at Hamadam, in Persia, 1036. Arab physician and naturalist. Edited several mathematical and physical works of Aristotle, Euclid, etc. Wrote on arithmetic and geometry.

Babbage, Charles. Born at Totnes, Dec. 26, 1792; died at London, Oct. 18, 1871. Lucasian professor of mathematics at Cambridge. Popularly known for his calculating machine. Did much to raise the standard of mathematics in England.

Bachet. See Méziriac.

Bacon, Roger. Born at Ilchester, Somersetshire, 1214; died at Oxford, June 11, 1294. Studied at Oxford and Paris; professor at Oxford; mathematician and physicist.

Balbus. Lived about 100. Roman surveyor.

Baldi, Bernardino. Born at Urbino, 1553; died there, 1617. Mathematician and general scholar. Contributed to the history of mathematics.

Baltzer, Heinrich Richard. Born at Meissen in 1818; died at Giessen in 1887. Professor of mathematics at Giessen.

Barlaam, Bernard. Beginning of fourteenth century. A monk who wrote on astronomy and geometry.

Barozzi, Francesco. Italian mathematician. 1537–1604.

Barrow, Isaac. Born at London, 1630; died at Cambridge, May 4, 1677. Professor of Greek and mathematics at Cambridge. Scholar, mathematician, scientist, preacher. Newton was his pupil and successor.

Beda, the Venerable. Born at Monkton, near Yarrow, Northumberland, in 672; died at Yarrow, May 26, 735. Wrote on chronology and arithmetic.

Bellavitis, Giusto. Born at Bassano, near Padua, Nov. 22, 1803; died Nov. 6, 1880. Known for his work in projective geometry and his method of equipollences.

Bernelinus. Lived about 1020. Pupil of Gerbert at Paris. Wrote on arithmetic.

Bernoulli. Famous mathematical family.

Jacob (often called James, by the English), born at Basel, Dec. 27, 1654; died there Aug. 16, 1705. Among the first to recognize the value of the calculus. His *De Arte Conjectandi* is a classic on probabilities. Prominent in the study of curves, the logarithmic spiral being engraved on his monument at Basel.

John (Johann), his brother; born at Basel, Aug. 7, 1667; died there Jan. 1, 1748. Made the first attempt to construct an integral and an exponential calculus. Also prominent as a physicist, but his abilities were chiefly as a teacher.

Nicholas (Nikolaus), his nephew; born at Basel, Oct. 10, 1687; died there Nov. 29, 1759. Professor at St. Petersburg, Basel, and Padua. Contributed to the study of differential equations.

Daniel, son of John; born at Gröningen, Feb. 9, 1700; died at Basel in 1782. Professor of mathematics at St. Petersburg. His chief work was on hydrodynamics.

John the younger, son of John. 1710-1790. Professor at Basel.

Bézout, Étienne. Born at Nemours in 1730; died at Paris in 1783. Algebraist, prominent in the study of symmetric functions and determinants.

Bhaskara Acharya. Born in 1114. Hindu mathematician and astronomer. Author of the *Lilavati* and the *Vijaganita,* containing the elements of arithmetic and algebra. One of the most prominent mathematicians of his time.

Biot, Jean Baptiste. Born at Paris, Apr. 21, **1774**; died same place Feb. 3, 1862. Professor of physics, mathematics, astronomy. Voluminous writer.

Boëthius, Anicius Manlius Torquatus Severinus. Born at Rome, 480; executed at Pavia, 524. Founder of medieval scholasticism. Translated and revised many Greek writings on mathematics, mechanics, and physics. Wrote on arithmetic. While in prison he composed his Consolations of Philosophy.

Bolyai: Wolfgang Bolyai de Bolya. Born at Bolya, **1775**; died in 1856. Friend of Gauss.

Johann Bolyai de Bolya, his son. Born at Klausenburg, 1802; died at Maros-Vàsàrhely, 1860. One of the discoverers (see Lobachevsky) of the so-called non-Euclidean geometry.

Bolzano, Bernhard. 1781-1848. Contributed to the study of series.

Bombelli, Rafaele. Italian. Born c. 1530. His algebra (1572) summarized all then known on the subject. Contributed to the study of the cubic.

Boncompagni, Baldassare. Wealthy Italian prince. Born at Rome. May 10, 1821; died at same place, April 12, 1894. Publisher of Boncompagni's Bulletino.

Boole, George. Born at Lincoln, 1815; died at Cork, 1864. Professor of mathematics in Queen's College, Cork. The theory of invariants and covariants may be said to start with his contributions (1841).

Booth, James. 1806-1878. Clergyman and writer on elliptic integrals.

Borchardt, Karl Wilhelm. Born in 1817; died at Berlin, 1880. Professor at Berlin.

Boschi, Pietro. Born at Rome, 1833; died in 1887. Professor at Bologna.

Bouquet, Jean Claude. Born at Morteau in 1819; died at Paris, 1885.

Bour, Jacques Edmond Émile. Born in 1832; died at Paris, 1866. Professor in the École Polytechnique.

Bradwardine, Thomas de. Born at Hardfield, near Chichester. 1290; died at Lambeth, Aug. 26, 1349. Professor of theology at Oxford and later Archbishop of Canterbury. Wrote upon arithmetic and geometry.

Brahmagupta. Born in 598. Hindu mathematician. Contributed to geometry and trigonometry.

Brasseur, Jean Baptiste. 1802-1868. Professor at Liège.

Bretschneider, Carl Anton. Born at Schneeberg, May 27, 1808, died at Gotha, November 6, 1878.

Brianchon, Charles Julien. Born at Sèvres, 1785; died in 1864. Celebrated for his reciprocal (1806) to Pascal's mystic hexagram

Briggs, Henry. Born at Warley Wood, near Halifax, Yorkshire, Feb. 1560-1; died at Oxford Jan. 26, 1630-1. Savilian Professor of geometry at Oxford. Among the first to recognize the value of logarithms; those with decimal base bear his name.

Briot, Charles August Albert. Born at Sainte-Hippolyte, 1817; died in 1882. Professor at the Sorbonne, Paris.

Brouncker, William, Lord. Born in 1620 (?); died at Westminster, 1684. First president of the Royal Society. Contributed to the theory of series.

Brunelleschi, Filippo. Born at Florence, 1379; died there April 16, 1446. Noted Italian architect.

Bürgi, Joost (Jobst). Born at Lichtensteig, St. Gall, Switzerland, 1552; died at Cassel in 1632. One of the first to suggest a system of logarithms. The first to recognize the value of making the second member of an equation zero.

Caporali, Ettore. Born at Perugia, 1855; died at Naples, 1886. Professor of mathematics and writer on geometry.

Cardan, Jerome (Hieronymus, Girolamo). Born at Pavia, 1501; died at Rome, 1576. Professor of mathematics at Bologna and Padua. Mathematician, physician, astrologer. Chief contributions to algebra and theory of epicycloids.

Carnot, Lazare Nicolas Marguerite. Born at Nolay, Côte d'Or, 1753; died in exile at Magdeburg, 1823. Contributed to modern geometry.

Cassini, Giovanni Domenico. Born at Perinaldo, near Nice, 1625; died at Paris, 1712. Professor of astronomy at Bologna, and first of the family which for four generations held the post of director of the observatory at Paris.

Castigliano, Carlo Alberto. 1847-1884 Italian engineer.

Catalan, Eugène Charles. Born at Bruges, Belgium, May 30, 1814; died Feb. 14, 1894. Professor of mathematics at Paris and Liège.

Cataldi, Pietro Antonio. Italian mathematician, born 1548; died at Bologna, 1626. Professor of mathematics at Florence, Perugia and Bologna. Pioneer in the use of continued fractions.

Cattaneo, Francesco. 1811-1875. Professor of physics and mechanics in the University of Pavia.

Cauchy, Augustin Louis. Born at Paris, 1789; died at Sceaux, 1857. Professor of mathematics at Paris. One of the most prominent mathematicians of his time. Contributed to the theory of functions, determinants, differential equations, theory of residues, elliptic functions, convergent series, etc.

Cavalieri, Bonaventura. Born at Milan, 1598; died at Bologna, 1647. Paved the way for the differential calculus by his method of indivisibles (1629).

Cayley, Arthur. Born at Richmond, Surrey, Aug. 16, 1821; died at Cambridge, Jan. 26, 1895. Sadlerian professor of mathematics, University of Cambridge. Prolific writer on mathematics.

Ceva, Giovanni. 1648–c. 1737. Contributed to the theory of transversals.

Chasles, Michel. Born at Chartres, Nov. 15, 1793; died at Paris, Dec. 12, 1880. Contributed extensively to the theory of modern geometry.

Chelini, Domenico. Born 1802; died Nov. 16, 1878. Italian mathematician; contributed to analytic geometry and mechanics.

Chuquet, Nicolas. From Lyons; died about 1500. Lived in Paris and contributed to algebra and arithmetic.

Clairaut, Alexis Claude. Born at Paris, 1713; died there, 1765. Physicist, astronomer, mathematician. Prominent in the study of curves.

Clausberg, Christlieb von. Born at Danzig, 1689; died at Copenhagen, 1751.

Clebsch, Rudolf Friedrich Alfred. Born January 19, 1833; died Nov. 7, 1872. Professor of mathematics at Carlsruhe, Giessen and Göttingen.

Condorcet, Marie Jean Antoine Nicolas. Born at Ribemont, near St. Quentin, Aisne, 1743; died at Bourg-la Reine, 1794. Secretary of the Académie des Sciences. Contributed to the theory of probabilities.

Cotes, Roger. Born at Burbage, near Leicester, July 10, 1682; died at Cambridge, June 5, 1716. Professor of astronomy at Cambridge. His name attaches to a number of theorems in geometry, algebra and analysis. Newton remarked, "If Cotes had lived we should have learnt something."

Cramer, Gabriel. Born at Geneva, 1704 ; diéd at Bagnols, 1752. Added to the theory of equations and revived the study of determinants (begun by Leibnitz). Wrote a treatise on curves.

Crelle, August Leopold. Born at Eichwerder (Wriezen a. d. Oder), 1780 ; died in 1855. Founder cf the *Journal für reine und angewandte Mathematik* (1826).

D'Alembert, Jean le Rond. Born at Paris, 1717 ; died there, 1783. Physicist, mathematician, astronomer. Contributed to the theory of equations.

De Beaune, Florimond. 1601-1652. Commentator on Descartes's Geometry.

De la Gournerie, Jules Antoine René Maillard. Born in 1814 ; died at Paris, 1833. Contributed to descriptive geometry.

Del Monte, Guidobaldo. 1545-1607. Wrote on mechanics and perspective.

Democritus. Born at Abdera, Thrace, —460 ; died *c.* —370. Studied in Egypt and Persia. Wrote on the theory of numbers and on geometry. Suggested the idea of the infinitesimal.

De Moivre, Abraham. Born at Vitry, Champagne, 1667 ; died at London, 1754. Contributed to the theory of complex numbers and of probabilities

De Morgan, Augustus. Born at Madura, Madras, June 1806 ; died March 18, 1871. First professor of mathematics in University of London (1828). Celebrated teacher, but also contributed to algebra and the theory of probabilities.

Desargues, Gerard. Born at Lyons, 1593 ; died in 1662. One of the founders of modern geometry.

Descartes, René, du Perron. Born at La Haye, Touraine, 1596 ; died at Stockholm, 1650. Discoverer of analytic geometry. Contributed extensively to algebra.

Dinostratus. Lived about —335. Greek geometer. Brother of Menaechmus. His name is connected with the quadratrix.

Diocles. Lived about —180. Greek geometer. Discovered the cissoid which he used in solving the Delian problem.

Diophantus of Alexandria. Lived about 275. Most prominent of Greek algebraists, contributing especially to indeterminate equations.

Dirichlet, Peter Gustav Lejeune. Born at Düren, 1805; died at Göttingen, 1859. Succeeded Gauss as professor at Göttingen Prominent contributor to the theory of numbers.

Dodson, James. Died Nov. 23, 1757. Great grandfather of De Morgan. Known chiefly for his extensive table of anti-logarithms (1742).

Donatello, 1386–1468. Italian sculptor.

Du Bois-Reymond, Paul David Gustav. Born at Berlin, Dec. 2, 1831; died at Freiburg, April 7, 1889. Professor of mathematics in Heidelberg, Freiburg, and Tübingen.

Duhamel, Jean Marie Constant. Born at Saint-Malo, 1797; died at Paris, 1872. One of the first to write upon method in mathematics.

Dupin, François Pierre Charles. Born at Varzy, 1784; died at Paris, 1873.

Dürer, Albrecht. Born at Nuremberg, 1471; died there, 1528. Famous artist. One of the founders of the modern theory of curves.

Eisenstein, Ferdinand Gotthold Max. Born at Berlin, 1823; died there, 1852. One of the earliest workers in the field of invariants and covariants.

Enneper, Alfred. 1830–1885. Professor at Göttingen.

Epaphroditus. Lived about 200. Roman surveyor. Wrote on surveying, theory of numbers, and mensuration.

Eratosthenes. Born at Cyrene, Africa, —276; died at Alexandria, —194. Prominent geographer. Known for his "sieve" for finding primes.

Euclid. Lived about —300. Taught at Alexandria in the reign of Ptolemy Soter. The author or compiler of the most famous text-book of Geometry ever written, the *Elements*, in thirteen books.

Eudoxus of Cnidus. —408, —355. Pupil of Archytas and Plato. Prominent geometer, contributing especially to the theories of proportion, similarity, and "the golden section."

Euler, Leonhard. Born at Basel, 1707; died at St. Petersburg, 1783. One of the greatest physicists, astronomers and mathematicians of the 18th century. "In his voluminous . .

writings will be found a perfect storehouse of investigations on every branch of algebraical and mechanical science."—Kelland.

Eutocius. Born at Ascalon, 480. Geometer. Wrote commentaries on the works of Archimedes, Apollonius, and Ptolemy.

Fagnano, Giulio Carlo, Count de. Born at Sinigaglia, 1682; died in 1766. Contributed to the study of curves. Euler credits him with the first work in elliptic functions.

Faulhaber, Johann. 1580-1635. Contributed to the theory of series.

Fermat, Pierre de. Born at Beaumont-de-Lomagne, near Montauban, 1601; died at Castres, Jan. 12, 1665. One of the most versatile mathematicians of his time; his work on the theory of numbers has never been equalled.

Ferrari, Ludovico. Born at Bologna, 1522; died in 1562. Solved the biquadratic.

Ferro, Scipione del. Born at Bologna, *c.* 1465; died between Oct. 29 and Nov. 16, 1526. Professor of mathematics at Bologna. Investigated the geometry based on a single setting of the compasses, and was the first to solve the special cubic $x^3 + px = q$.

Feuerbach, Karl Wilhelm. Born at Jena, 1800; died in 1834. Contributed to modern elementary geometry.

Fibonacci. See Leonardo of Pisa.

Fourier, Jean Baptiste Joseph, Baron. Born at Auxerre, 1768; died at Paris, 1830. Physicist and mathematician. Contributed to the theories of equations and of series.

Frénicle. Bernard *Frénicle* de Bessy. 1605-1675. Friend of Fermat.

Frézier, Amédée François. Born at Chambéry, 1682; died at Brest, 1773. One of the founders of descriptive geometry.

Friedlein, Johann Gottfried. Born at Regensburg, 1828; died in 1875.

Frontinus, Sextus Julius. 40-103. Roman surveyor and engineer.

Galois, Evariste. Born at Paris, 1811; died there, 1832. Founder of the theory of groups.

Gauss, Karl Friedrich. Born at Brunswick, 1777; died at Göttingen, 1855. The greatest mathematician of modern times. Prominent as a physicist and astronomer. The theories of numbers, of functions, of equations, of determinants, of complex numbers, of hyperbolic geometry, are all largely indebted to his great genius.

Geber. Jabir ben Aflah. Lived about 1085. Astronomer at Seville; wrote on spherical trigonometry.

Gellibrand, Henry. 1597–1637. Professor of astronomy at Gresham College.

Geminus. Born at Rhodes, —100; died at Rome, —40. Wrote on astronomy and (probably) on the history of pre-Euclidean mathematics.

Gerbert, Pope Sylvester II. Born at Auvergne, 940; died at Rome, May 13, 1003. Celebrated teacher; elected pope in 999. Wrote upon arithmetic.

Gerhard of Cremona. From Cremona (or, according to others, Carmona in Andalusia). Born in 1114; died at Toledo in 1187. Physician, mathematician, and astrologer. Translated several works of the Greek and Arab mathematicians from Arabic into Latin.

Germain, Sophie. 1776–1831. Wrote on elastic surfaces.

Girard, Albert. *c.* 1590–1633. Contributed to the theory of equations, general polygons, and symbolism.

Göpel, Gustav Adolf. 1812–1847. Known for his researches on hyperelliptic functions.

Grammateus, Henricus. (German name, Heinrich Schreiber.) Born at Erfurt, *c.* 1476. Arithmetician.

Grassmann, Hermann Günther. Born at Stettin, April 15, 1809; died there Sept. 26, 1877. Chiefly known for his *Ausdehnungslehre* (1844). Also wrote on arithmetic, trigonometry, and physics.

Grebe, Ernst Wilhelm. Born near Marbach, Oberhesse, Aug. 30, 1804; died at Cassel, Jan. 14, 1874. Contributed to modern elementary geometry.

Gregory, James. Born at Drumoak, Aberdeenshire, Nov. 1638; died at Edinburgh, 1675. Professor of mathematics at St. Andrews and Edinburgh. Proved the incommensurability of π; contributed to the theory of series.

Grunert, Johann August. Born at Halle a. S., 1797; died in 1872. Professor at Greifswalde, and editor of Grunert's *Archiv*.

Gua. Jean Paul de *Gua* de Malves. Born at Carcassonne, 1713; died at Paris, June 2, 1785. Gave the first rigid proof of Descartes's rule of signs.

Gudermann, Christoph. Born at Winneburg, March 28, 1798; died at Münster, Sept. 25, 1852. To him is largely due the introduction of hyperbolic functions into modern analysis.

Guldin, Habakkuk (Paul). Born at St. Gall, 1577; died at Grätz, 1643. Known chiefly for his theorem on a solid of revolution, pilfered from Pappus.

Hachette, Jean Nicolas Pierre. Born at Mézières, 1769; died at Paris, 1834. Algebraist and geometer.

Halley, Edmund. Born at Haggerston, near London, Nov. 8, 1656; died at Greenwich, Jan. 14, 1742. Chiefly known for his valuable contributions to physics and astronomy.

Halphen, George Henri. Born at Rouen, Oct. 30, 1844; died at Versailles in 1889. Professor in the École Polytechnique at Paris. Contributed to the theories of differential equations and of elliptic functions.

Hamilton, Sir William Rowan. Born at Dublin, Aug. 3–4, 1805; died there, Sept. 2, 1865. Professor of astronomy at Dublin. Contributed extensively to the theory of light and to dynamics, but known generally for his discovery of quaternions.

Hankel, Hermann. Born at Halle, Feb. 14, 1839; died at Schramberg, Aug. 29, 1873. Contributed chiefly to the theory of complex numbers and to the history of mathematics.

Harnack, Karl Gustav Axel. Born at Dorpat, 1851; died at Dresden in 1888. Professor in the polytechnic school at Dresden.

Harriot, Thomas. Born at Oxford, 1560; died at Sion House, near Isleworth, July 2, 1621. The most celebrated English algebraist of his time.

Heron of Alexandria. Lived about —110. Celebrated surveyor and mechanician. Contributed to mensuration.

Hesse, Ludwig Otto. Born at Königsberg, April 22, 1811; died
at Munich, Aug. 4, 1874. Contributed to the theories of curves
and of determinants.

Hipparchus. Born at Nicaea, Bithynia, —180; died at Rhodes,
—125. Celebrated astronomer. One of the earliest writers
on spherical trigonometry.

Hippias of Elis. Born *c.* —460. Mathematician, astronomer,
natural scientist. Discovered the quadratrix.

Hippocrates of Chios. Lived about —440. Wrote the first Greek
elementary text-book on mathematics.

Horner, William George. Born in 1786; died at Bath, Sept. 22,
1837. Chiefly known for his method of approximating the real
roots of a numerical equation (1819).

Hrabanus Maurus. 788–856. Teacher of mathematics. Arch
bishop of Mainz.

Hudde, Johann. Born at Amsterdam, 1633; died there, 1704.
Contributed to the theories of equations and of series.

Honein ibn Ishak. Died in 873. Arab physician. Translated
several Greek scientific works.

Huygens, Christiaan, van Zuylichem. Born at the Hague, 1629;
died there, 1695. Famous physicist and astronomer. In mathematics he contributed to the study of curves.

Hyginus. Lived about 100. Roman surveyor.

Hypatia, daughter of Theon of Alexandria. 375–415. Composed
several mathematical works. See Charles Kingsley's *Hypatia.*

Hypsicles of Alexandria. Lived about —190. Wrote on solid
geometry and theory of numbers, and solved certain indeterminate equations.

Iamblichus. Lived about 325. From Chalcis. Wrote on various
branches of mathematics.

Ibn al Banna. Abul Abbas Ahmed ibn Mohammed ibn Otman al
Azdi al Marrakushi ibn al Banna Algarnati. Born 1252 or
1257 in Morocco. West Arab algebraist; prolific writer.

Ibn Yunus, Abul Hasan Ali ibn Abi Said Abderrahman. 960–
1008. Arab astronomer; prepared the Hakimitic Tables.

Isidorus Hispalensis. Born at Carthagena, 570; died at Seville, 636. Bishop of Seville. His *Origines* contained dissertations on mathematics.

Ivory, James. Born at Dundee, 1765; died at London, Sept. 21, 1842. Chiefly known as a physicist.

Jacobi, Karl Gustav Jacob. Born at Potsdam, Dec. 10, 1804; died at Berlin, Feb. 18, 1851. Important contributor to the theory of elliptic and theta functions and to that of functional determinants.

Jamin, Jules Célestin. Born in 1818; died at Paris, 1886. Professor of physics.

Joannes de Praga (Johannes Schindel). Born at Königgrätz, 1370 or 1375; died at Prag *c.* 1450. Astronomer and mathematician.

Johannes of Seville (Johannes von Luna, Johannes Hispalensis). Lived about 1140. A Spanish Jew; wrote on arithmetic and algebra.

Johann von Gmünden. Born at Gmünden am Trannsee, between 1375 and 1385; died at Vienna, Feb. 23, 1442. Professor of mathematics and astronomy at Vienna; the first full professor of mathematics in a Teutonic university.

Kästner, Abraham Gotthelf. Born at Leipzig, 1719; died at Göttingen, 1800. Wrote on the history of mathematics.

Kepler, Johann. Born in Würtemberg, near Stuttgart, 1571; died at Regensburg, 1630. Astronomer (assistant of Tycho Brahe, as a young man); "may be said to have constructed the edifice of the universe,"—Proctor. Prominent in introducing the use of logarithms. Laid down the "principle of continuity" (1604); helped to lay the foundation of the infinitesimal calculus.

Khayyam, Omar. Died at Nishapur, 1123. Astronomer, geometer, algebraist. Popularly known for his famous collection of quatrains, the *Rubaiyat.*

Köbel, Jacob. Born at Heidelberg, 1470; died at Oppenheim, in 1533. Prominent writer on arithmetic (1514, 1520).

Lacroix, Sylvestre François. Born at Paris, 1765; died there, May 25, 1843. Author of an elaborate course of mathematics.

Laguerre, Edmond Nicolas. Born at Bar-le-Duc, April 9, 1834; died there Aug. 14, 1886. Contributed to higher analysis.

Lagrange, Joseph Louis, Comte. Born at Turin, Jan. 25, 1736; died at Paris, April 10, 1813. One of the foremost mathematicians of his time. Contributed extensively to the calculus of variations, theory of numbers, determinants, differential equations, calculus of finite differences, theory of equations, and elliptic functions. Author of the *Mécanique analytique*. Also celebrated as an astronomer.

Lahire, Philippe de. Born at Paris, March 18, 1640; died there April 21, 1718. Contributed to the study of curves and magic squares.

Laloubère, Antoine de. Born in Languedoc, 1600; died at Toulouse, 1664. Contributed to the study of curves.

Lambert, Johann Heinrich. Born at Mülhausen, Upper Alsace, 1728; died at Berlin, 1777. Founder of the hyperbolic trigonometry.

Lamé, Gabriel. Born at Tours, 1795; died at Paris, 1870. Writer on elasticity and orthogonal surfaces.

Landen, John. Born at Peakirk, near Peterborough. 1719; died at Milton, 1790. A theorem of his (1755) suggested to Euler and Lagrange their study of elliptic integrals.

Laplace, Pierre Simon, Marquis de. Born at Beaumont-en-Auge, Normandy, March 23, 1749; died at Paris, March 5, 1827. Celebrated astronomer, physicist, and mathematician. Added to the theories of least squares, determinants, equations, series, probabilities, and differential equations.

Legendre, Adrien Marie. Born at Toulouse, Sept. 18, 1752; died at Paris, Jan. 10, 1833. Celebrated mathematician, contributing especially to the theory of elliptic functions, theory of numbers, least squares, and geometry. Discovered the "law of quadratic reciprocity,"—"the gem of arithmetic" (Gauss).

Leibnitz, Gottfried Wilhelm. Born at Leipzig, 1646; died at Hanover in 1716. One of the broadest scholars of modern times; equally eminent as a philosopher and mathematician. One of the discoverers of the infinitesimal calculus, and the inventor of its accepted symbolism.

Leonardo of Pisa, Fibonacci (filius Bonacii, son of Bonacius). Born at Pisa, 1180; died in 1250. Travelled extensively and brought back to Italy a knowledge of the Hindu numerals and the general learning of the Arabs, which he set forth in his *Liber Abaci, Practica geometriae*, and *Flos*.

L'Hospital, Guillaume François Antoine de, Marquis de St. Mesme. Born at Paris, 1661; died there 1704. One of the first to recognise the value of the infinitesimal calculus.

Lhuilier, Simon Antoine Jean. Born at Geneva, 1750; died in 1840. Geometer.

Libri, Carucci dalla Sommaja, Guglielmo Brutus Icilius Timoleon. Born at Florence, Jan. 2, 1803; died at Villa Fiesole, Sept. 28, 1869. Wrote on the history of mathematics in Italy.

Lie, Marius Sophus. Born Dec. 12, 1842; died Feb. 18, 1899. Professor of mathematics in Christiania and Leipzig. Specially celebrated for his theory of continuous groups of transformations as applied to differential equations.

Liouville, Joseph. Born at St. Omer, 1809; died in 1882. Founder of the journal that bears his name.

Lobachevsky, Nicolai Ivanovich. Born at Makarief, 1793; died at Kasan, Feb. 12-24, 1856. One of the founders of the so-called non-Euclidean geometry.

Ludolph van Ceulen. See Van Ceulen.

MacCullagh, James. Born near Strabane, 1809; died at Dublin, 1846. Professor of mathematics and physics in Trinity College, Dublin.

Maclaurin, Colin. Born at Kilmodan, Argyllshire, 1698; died at York, June 14, 1746. Professor of mathematics at Edinburgh. Contributed to the study of conics and series

Malfatti, Giovanni Francesco Giuseppe. Born at Ala, Sept. 26, 1731; died at Ferrara, Oct. 9, 1807. Known for the geometric problem which bears his name,

Malus, Étienne Louis. Born at Paris, June 23, 1775; died there, Feb. 24, 1812. Physicist.

Mascheroni, Lorenzo. Born at Castagneta, 1750; died at Paris, 1800. First to elaborate the geometry of the compasses only (1795).

Maurolico, Francesco. Born at Messina, Sept. 16, 1494; died July 21, 1575. The leading geometer of his time. Wrote also on trigonometry.

Maximus Planudes. Lived about 1330. From Nicomedia. Greek mathematician at Constantinople. Wrote a commentary on Diophantus; also on arithmetic.

Menaechmus. Lived about —350. Pupil of Plato. Discoverer of the conic sections.

Menelaus of Alexandria. Lived about 100. Greek mathematician and astronomer. Wrote on geometry and trigonometry.

Mercator, Gerhard. Born at Rupelmonde, Flanders, 1512; died at Duisburg, 1594. Geographer.

Mercator, Nicholas. (German name Kaufmann.) Born near Cismar, Holstein, *c.* 1620; died at Paris, 1687. Discovered the series for log $(1 + x)$.

Metius, Adriaan. Born at Alkmaar, 1571; died at Franeker, 1635. Suggested an approximation for π, really due to his father.

Meusnier de la Place, Jean Baptiste Marie Charles. Born at Paris, 1754; died at Cassel, 1793. Contributed a theorem on the curvature of surfaces.

Méziriac, Claude Gaspard Bachet de. Born at Bourg-en-Bresse, 1581; died in 1638. Known for his *Problèmes plaisants*, etc. (1624) and his translation of Diophantus.

Möbius, August Ferdinand. Born at Schulpforta, Nov. 17, 1790; died at Leipzig, Sept. 26, 1868. One of the leaders in modern geometry. Author of *Der Barycentrische Calcül* (1827).

Mohammed ibn Musa. See Al Khowarazmi.

Moivre. See DeMoivre.

Mollweide, Karl Brandan. Born at Wolfenbüttel, Feb. 3, 1774; died at Leipzig, March 10, 1825. Wrote on astronomy and mathematics.

Monge, Gaspard, Comte de Péluse. Born at Beaune, 1746; died at Paris, 1818. Discoverer of descriptive geometry; contributed to the study of curves and surfaces, and to differential equations.

Montmort, Pierre Rémond de. Born at Paris, 1678; died there, 1719. Contributed to the theory of probabilities and to the summation of series.

Moschopulus, Manuel. Lived about 1300. Byzantine mathematician. Known for his work on magic squares.

Mydorge, Claude. Born at Paris, 1585; died there in 1647. Author of the first French treatise on conics.

Napier, John. Born at Merchiston, then a suburb of Edinburgh, 1550; died there in 1617. Inventor of logarithms. Contributed to trigonometry.

Newton, Sir Isaac. Born at Woolsthorpe, Lincolnshire, Dec. 25, 1642, O. S.; died at Kensington, March 20, 1727. Succeeded Barrow as Lucasian professor of mathematics at Cambridge (1669). The world's greatest mathematical physicist. Invented fluxional calculus (*c.* 1666). Contributed extensively to the theories of series, equations, curves, and, in general, to all branches of mathematics then known.

Nicole, François. Born at Paris, 1683; died there, 1758. First treatise on finite differences.

Nicomachus of Gerasa, Arabia. Lived 100. Wrote upon arithmetic.

Nicomedes of Gerasa. Lived —180. Discovered the conchoid which bears his name.

Nicolaus von Cusa. Born at Cuss on the Mosel, 1401; died at Todi, Aug. 11, 1464. Theologian, physicist, astronomer, geometer.

Odo of Cluny. Born at Tours, 879; died at Cluny, 942 or 943. Wrote on arithmetic.

Oenopides of Chios. Lived —465. Studied in Egypt. Geometer.

Olivier, Théodore. Born at Lyons, Jan. 21, 1793; died in same place Aug. 5, 1853. Writer on descriptive geometry.

Oresme, Nicole. Born in Normandy, *c.* 1320; died at Lisieux, 1382. Wrote on arithmetic and geometry.

Oughtred, William. Born at Eton, 1574; died at Albury, 1660. Writer on arithmetic and trigonometry.

Pacioli, Luca. Fra Luca di Borgo di Santi Sepulchri. Born at Borgo San Sepolcro, Tuscany, *c.* 1445; died at Florence, *c.* 1509. Taught in several Italian cities. His *Summa de Arithmetica, Geometria*, etc., was the first great mathematical work published (1494).

Pappus of Alexandria. Lived about 300. Compiled a work containing the mathematical knowledge of his time.

Parent, Antoine. Born at Paris, 1665; died there in 1716. First to refer a surface to three co-ordinate planes (1700).

Pascal, Blaise. Born at Clermont, 1623; died at Paris, 1662. Physicist, philosopher, mathematician. Contributed to the theory of numbers, probabilities, and geometry.

Peirce, Charles S. Born at Cambridge, Mass., Sept. 10, 1839. Writer on logic.

Pell, John. Born in Sussex, March 1, 1610; died at London, Dec 10, 1685. Translated Rahn's algebra.

Perseus. Lived —150. Greek geometer; studied spiric lines.

Peuerbach, Georg von. Born at Peuerbach, Upper Austria, May 30, 1423; died at Vienna, April 8, 1461. Prominent teacher and writer on arithmetic, trigonometry, and astronomy.

Pfaff, Johann Friedrich. Born at Stuttgart, 1765; died at Halle in 1825. Astronomer and mathematician.

Pitiscus, Bartholomaeus. Born Aug 24, 1561; died at Heidelberg, July 2, 1613. Wrote on trigonometry, and first used the present decimal point (1612).

Plana, Giovanni Antonio Amedeo. Born at Voghera, Nov. 8, 1781; died at Turin, Jan. 2, 1864. Mathematical astronomer and physicist.

Planudes. See Maximus Planudes.

Plateau, Joseph Antoine Ferdinand. Born at Brussels, Oct. 14, 1801; died at Ghent, Sept. 15, 1883. Professor of physics at Ghent.

Plato. Born at Athens, —429; died in —348. Founder of the Academy. Contributed to the philosophy of mathematics.

Plato of Tivoli. Lived 1120. Translated Al Battani's trigonometry and other works.

Plücker, Johann. Born at Elberfeld, July 16, 1801; died at Bonn, May 22, 1868. Professor of mathematics at Bonn and Halle. One of the foremost geometers of the century.

Poisson, Siméon Denis. Born at Pithiviers, Loiret, 1781 ; died at Paris, 1840. Chiefly known as a physicist. Contributed to the study of definite integrals and of series.

Poncelet, Jean Victor. Born at Metz, 1788 ; died at Paris, 1867. One of the founders of projective geometry.

Pothenot, Laurent. Died at Paris in 1732. Professor of mathematics in the Collège Royale de France.

Proclus. Born at Byzantium, 412 ; died in 485. Wrote a commentary on Euclid. Studied higher plane curves.

Ptolemy (Ptolemaeus Claudius). Born at Ptolemais, 87 ; died at Alexandria, 165. One of the greatest Greek astronomers.

Pythagoras. Born at Samos, —580 ; died at Megapontum, —501. Studied in Egypt and the East. Founded the Pythagorean school at Croton, Southern Italy. Beginning of the theory of numbers. Celebrated geometrician.

Quetelet, Lambert Adolph Jacques. Born at Ghent, Feb. 22, 1796 ; died at Brussels, Feb. 7, 1874. Director of the royal observatory of Belgium. Contributed to geometry, astronomy, and statistics.

Ramus, Peter (Pierre de la Ramée). Born at Cuth, Picardy, 1515 ; murdered at the massacre of St. Bartholomew, Paris, August 24–25, 1572. Philosopher, but also a prominent writer on mathematics.

Recorde, Robert. Born at Tenby, Wales, 1510 ; died in prison, at London, 1558. Professor of mathematics and rhetoric at Oxford. Introduced the sign = for equality.

Regiomontanus. Johannes Müller. Born near Königsberg, June 6, 1436 ; died at Rome, July 6, 1476. Mathematician, astronomer, geographer. Translator of Greek mathematics. Author of first text-book of trigonometry.

Remigius of Auxerre. Died about 908. Pupil of Alcuin's. Wrote on arithmetic.

Rhaeticus, Georg Joachim. Born at Feldkirch, 1514 ; died at Kaschau, 1576. Professor of mathematics at Wittenberg ; pupil of Copernicus and editor of his works. Contributed to trigonometry.

Riccati, Count Jacopo Francesco. Born at Venice, 1676; died at Trèves, 1754. Contributed to physics and differential equations.

Richelot, Friedrich Julius. Born at Königsberg, Nov. 6, 1808; died March 31, 1875 in same place. Wrote on elliptic and Abelian functions.

Riemann, George Friedrich Bernhard. Born at Breselenz, Sept. 17, 1826; died at Selasca, July 20, 1866. Contributed to the theory of functions and to the study of surfaces.

Riese, Adam. Born at Staffelstein, near Lichtenfels, 1492; died at Annaberg, 1559. Most influential teacher of and writer on arithmetic in the 16th century.

Roberval, Giles Persone de. Born at Roberval, 1602; died at Paris, 1675. Professor of mathematics at Paris. Geometry of tangents and the cycloid.

Rolle, Michel. Born at Ambert, April 22, 1652; died at Paris, Nov. 8, 1719. Discovered the theorem which bears his name, in the theory of equations.

Rudolff, Christoff. Lived in first part of the sixteenth century. German algebraist.

Sacro-Bosco, Johannes de. Born at Holywood (Halifax), Yorkshire, 1200(?); died at Paris, 1256. Professor of mathematics and astronomy at Paris. Wrote on arithmetic and trigonometry.

Saint-Venant, Adhémar Jean Claude Barré de. Born in 1797; died in Vendôme, 1886. Writer on elasticity and torsion.

Saint-Vincent, Gregoire de. Born at Bruges, 1584; died at Ghent, 1667. Known for his vain attempts at circle squaring.

Saurin, Joseph. Born at Courtaison, 1659; died at Paris, 1737. Geometry of tangents.

Scheeffer, Ludwig. Born at Königsberg, 1859; died at Munich, 1885. Writer on theory of functions.

Schindel, Johannes. See Joannes de Praga.

Schwenter, Daniel. Born at Nuremberg, 1585; died in 1636. Professor of oriental languages and of mathematics at Altdorf.

Serenus of Antissa. Lived about 350. Geometer.

Serret, Joseph Alfred. Born at Paris, Aug. 30, 1819; died at Versailles, March 2, 1885. Author of well-known text-books on algebra and the differential and integral calculus.

Sextus Julius Africanus. Lived about 220. Wrote on the history of mathematics.

Simpson, Thomas. Born at Bosworth, Aug. 20, 1710; died at Woolwich, May 14, 1761. Author of text-books on algebra, geometry, trigonometry, and fluxions.

Sluze, René François Walter de. Born at Visé on the Maas, 1622; died at Liège in 1685. Contributed to the notation of the calculus, and to geometry.

Smith, Henry John Stephen. Born at Dublin, 1826; died at Oxford, Feb. 9, 1883. Leading English writer on theory of numbers.

Snell, Willebrord, van Roijen. Born at Leyden, 1591; died there, 1626. Physicist, astronomer, and contributor to trigonometry.

Spottiswoode, William. Born in London, Jan. 11, 1825; died there, June 27, 1883. President of the Royal Society. Writer on algebra and geometry.

Staudt, Karl Georg Christian von. Born at Rothenburg a. d. Tauber, Jan. 24, 1798; died at Erlangen, June 1, 1867. Prominent contributor to modern geometry, Geometrie der Lage.

Steiner, Jacob. Born at Utzendorf, March 18, 1796; died at Bern, April 1, 1863. Famous geometrician.

Stevin, Simon. Born at Bruges, 1548; died at Leyden (or the Hague), 1620. Physicist and arithmetician.

Stewart, Matthew. Born at Rothsay, Isle of Bute, 1717; died at Edinburgh, 1785. Succeeded Maclaurin as professor of mathematics at Edinburgh. Contributed to modern elementary geometry.

Stifel, Michael. Born at Esslingen, 1486 or 1487; died at Jena, 1567. Chiefly known for his *Arithmetica integra* (1544).

Sturm, Jacques Charles François. Born in Geneva, 1803; died in 1855. Professor in the École Polytechnique at Paris. "Sturm's theorem."

Sylvester, James Joseph. Born in London, Sept. 3, 1814; died in same place, March 15, 1897. Savilian professor of pure geometry in the University of Oxford. Writer on algebra, especially the theory of invariants and covariants.

Tabit ibn Kurra. Born at Harran in Mesopotamia, 833; died at Bagdad, 902. Mathematician and astronomer. Translated works of the Greek mathematicians, and wrote on the theory of numbers.

Tartaglia, Nicolo. (Nicholas the Stammerer. Real name, Nicolo Fontana.) Born at Brescia, *c.* 1500; died at Venice, *c.* 1557. Physicist and arithmetician; best known for his work on cubic equations.

Taylor, Brook. Born at Edmonton, 1685; died at London, 1731. Physicist and mathematician. Known chiefly for his work in series.

Thales. Born at Miletus, —640; died at Athens, —548. One of the "seven wise men" of Greece; founded the Ionian School. Traveled in Egypt and there learned astronomy and geometry. First scientific geometry in Greece.

Theaetetus of Heraclea. Lived in —390. Pupil of Socrates. Wrote on irrational numbers and on geometry.

Theodorus of Cyrene. Lived in —410. Plato's mathematical teacher. Wrote on irrational numbers.

Theon of Alexandria. Lived in 370. Teacher at Alexandria. Edited works of Greek mathematicians.

Theon of Smyrna Lived in 130. Platonic philosopher. Wrote on arithmetic, geometry, mathematical history, and astronomy.

Thymaridas of Paros. Lived in —390. Pythagorean; wrote on arithmetic and equations.

Torricelli, Evangelista. Born at Faënza, 1608; died in 1647. Famous physicist.

Tortolini, Barnaba. Born at Rome, Nov. 19, 1808; died August 24, 1874. Editor of the *Annali* which bear his name.

Trembley, Jean. Born at Geneva, 1749; died in 1811. Wrote on differential equations.

Tschirnhausen, Ehrenfried Walter, Graf von. Born at Kiesslingswalde, 1651; died at Dresden, 1708. Founded the theory of catacaustics.

Ubaldi, Guido. See Del Monte.

Unger, Ephraim Solomon. Born at Coswig, 1788; died in 1870.

Ursinus, Benjamin. 1587—1633. Wrote on trigonometry and computed tables.

Van Ceulen, Ludolph. Born at Hildesheim, Jan. 18 (or 28), 1540; died in Holland, Dec. 31, 1610. Known for his computations of π.

Vandermonde, Charles Auguste. Born at Paris, in 1735; died there, 1796. Director of the Conservatoire pour les arts et métiers.

Van Eyck, Jan. 1385–1440. Dutch painter.

Van Schooten, Franciscus (the younger). Born in 1615; died in 1660. Editor of Descartes and Vieta.

Vieta (Vieta), François, Seigneur de la Bigotière. Born at Fontenay-le-Comte, 1540; died at Paris, 1603. The foremost algebraist of his time. Also wrote on trigonometry and geometry.

Vincent. See Saint-Vincent.

Vitruvius. Marcus Vitruvius Pollio. Lived in —15. Roman architect. Wrote upon applied mathematics.

Viviani, Vincenzo. Born at Florence, 1622; died there, 1703. Pupil of Galileo and Torricelli. Contributed to elementary geometry.

Wallace, William. Born in 1768; died in 1843. Professor of mathematics at Edinburgh.

Wallis, John. Born at Ashford, 1616; died at Oxford, 1703. Savilian professor of geometry at Oxford. Published many mathematical works. Suggested (1685) the modern graphic interpretation of the imaginary.

Weierstrass, Karl Theodor Wilhelm. Born at Ostenfelde, Oct. 31, 1815; died at Berlin, Feb. 19, 1897. One of the ablest mathematicians of the century.

Werner, Johann. Born at Nuremberg, 1468; died in 1528. Wrote on mathematics, geography, and astronomy.

Widmann, Johann, von Eger. Lived in 1489. Lectured on algebra at Leipzig. The originator of German algebra. Wrote also on arithmetic and geometry.

Witt, Jan de.   Born in 1625, died in 1672.   Friend and helper of Descartes.

Wolf, Johann Christian von.   Born at Breslau, 1679; died at Halle, 1754.   Professor of mathematics and physics at Halle, and Marburg.   Text-book writer.

Woepcke, Franz.   Born at Dessau, May 6, 1826; died at Paris, March 25, 1864.   Studied the history of the development of mathematical sciences among the Arabs.

Wren, Sir Christopher.   Born at East Knoyle, 1632; died at London, in 1723.   Professor of astronomy at Gresham College; Savilian professor at Oxford; president of the Royal Society. Known, however, entirely for his great work as an architect

INDEX.*

* The numbers refer to pages, the small italic n's to footnotes.

www.ingramcontent.com/pod-product-compliance
Lightning Source LLC
Chambersburg PA
CBHW071710170526
45165CB00005B/1962